Vaccines and Diagnostics for Transboundary Animal Diseases

Developments in Biologicals

Vol. 135

This series "Developments in Biologicals" begins with Vol. 102 and is the continuation of IABS symposia series "Progress in Immunobiological Standardization", Vols 1-5, "Immunobiological Standardization", Vols 1-22 and "Developments in Biological Standardization", Vols 23-101.

Organized and published by the International Alliance for Biological Standardization (IABS)

KARGER

Basel . Freiburg . Paris . London . New York
Bangalore . Bangkok . Singapore . Tokyo . Sydney

Vaccines and Diagnostics for Transboundary Animal Diseases

Ames, Iowa
17-19 September 2012

Volume editors

James A. Roth, DVM, PhD, DACVM
Center for Food Security and Public Health, College of Veterinary Medicine, Iowa State University, Ames, IA, USA

Juergen A. Richt, DVM, PhD
Center of Excellence for Emerging and Zoonotic Animal Diseases (CEEZAD), College of Veterinary Medicine, Kansas State University, Manhattan, KS, USA

Igor A. Morozov, DVM, PhD
Center of Excellence for Emerging and Zoonotic Animal Diseases (CEEZAD), College of Veterinary Medicine, Kansas State University, Manhattan, KS, USA

Technical editors

Jane Galyon, MS
Center for Food Security and Public Health, College of Veterinary Medicine, Iowa State University, Ames, IA, USA

Betty Dodet, PharmD, PhD
IABS

Proceedings of an International Symposium organized by:

- US Department of Homeland Security, Science and Technology Directorate
- DHS Center of Excellence for Emerging and Zoonotic Animal Diseases (CEEZAD) at Kansas State University
- Center for Food Security and Public Health/Institute for International Cooperation in Animal Biologics at Iowa State University
- World Organisation for Animal Health (OIE)
- International Alliance for Biological Standardization (IABS)

14 figures and 16 tables, 2013

KARGER

Basel . Freiburg . Paris . London . New York
Bangalore . Bangkok . Singapore . Tokyo . Sydney

• • • • • • • • • • • • • • • • • •

Developments in Biologicals

Library of Congress Cataloging-in-Publication Data

Library of Congress Cataloging-in-Publication Data

Vaccines and Diagnostics for Transboundary Animal Diseases Workshop (2012 :
Ames, Iowa), author.
Vaccines and diagnostics for transboundary animal diseases : Ames, Iowa,
17-19 September 2012 / volume editors, James A. Roth, Jürgen A. Richt, Igor Morozov ; technical editors, Jane Galyon, Betty Dodet.
p. ; cm. -- (Developments in biologicals ; vol. 135)
"Proceedings of an International Symposium organized by: US Department of Homeland Security, Science and Technology Directorate [and 4 others]."
Includes bibliographical references and index.
ISBN 978-3-318-02365-7 (hard cover : alk. paper) -- ISBN 978-3-318-02366-4
(electronic version)
1. Communicable diseases in animals--Diagnosis. 2. Veterinary vaccines.
3. Veterinary clinical pathology. 4. Veterinary public health. I. Roth, James A., editor of compilation. II. Richt, J. A. (Jürgen Albrecht), editor of compilation. III. Morozov, Igor (Igor A.), editor of compilation. IV. United States. Department of Homeland Security. Science and Technology Directorate, sponsoring body. V. Title. VI. Series: Developments in biologicals ; v. 135.
[DNLM: 1. Animal Diseases--therapy--Congresses. 2. Animal
Diseases--diagnosis--Congresses. 3. Communicable Diseases,
Emerging--veterinary--Congresses. 4. Disease
Outbreaks--veterinary--Congresses. 5. Internationality--Congresses. 6.
Vaccines--therapeutic use--Congresses. W1 DE997VKF v.135 2013 / SF 740]
SF781.V33 2013
636.089'69--dc23

2013002143

Bibliographic Indices. This publication is listed in bibliographic services, including Current Contents® and Index Medicus.

Drug dosage. The authors and the publisher have exerted every effort to ensure that drug selection and dosage set forth in this text are in accord with current recommendations and practice at the time of publication. However, in view of ongoing research, changes in government regulations, and the constant flow of information relating to drug therapy and drug reactions, the reader is urged to check the package insert for each drug for any change in indications and dosage and for added warnings and precautions. This is particularly important when the recommended agent is a new and/or infrequently employed drug.

Printed in France by IML, Saint-Matin-en-haut (France). ISBN 978-3-318-02365-7

The scientific integrity of the articles contained in this volume is the sole responsibility of the authors.

Scientific Committee

James A. Roth, CFSPH, ISU/CVM, USA (co-chair)
Juergen A. Richt, CEEZAD, KSU/CVM, USA (co-chair)
Soren Alexandersen, NCFAD, Canada
Heinz Feldmann, NIH, USA
Cyril Gérard Gay, ARS/USDA, USA
Richard Hill, CVB, APHIS/USDA, USA
Beth Lautner, NVSL, APHIS/USDA, USA
Samia Metwally, FAO, Italy
Igor A.Morozov, CEEZAD, KSU/CVM, USA
Gerrit Viljoen, FAO/IAEA, Austria

DHS Liaison

Michelle M. Colby, DHS Science and Technology Directorate, Washington, DC, USA

Organization of the Proceedings

Preface

Session I	Roles and responsibilities – Facilitation and cooperation
Session II	State of the art, progress and gaps in development of vaccines and diagnostics for high priority transboundary animal diseases for the National Veterinary Stockpile

Acknowledgements

The Vaccines and Diagnostics for Transboundary Animal Diseases Workshop was made possible by financial support from the following organizations:

US Department of Homeland Security, Science and Technology Directorate, Grant Award Number 2010-ST-061-AG001

DHS Center of Excellence for Emerging & Zoonotic Animal Diseases (CEEZAD), Kansas State University, College of Veterinary Medicine, Manhattan, KS

Center for Food Security & Public Health, Institute for International Cooperation in Animal Biologics, College of Veterinary Medicine, Iowa State University, Ames, IA.

Industry Support:
Benchmark Biolabs
Boehringer Ingelheim Vetmedica, Inc.
Pfizer Animal Health
Merial Limited

Other Sponsorship:
World Organisation for Animal Health (OIE), Paris, France
International Alliance for Biological Standardization, Geneva, Switzerland

Contents

Roth JA, Richt JA, Morozov IA (eds): Vaccines and Diagnostics for Transboundary Animal Diseases. Dev Biol (Basel). Basel, Karger, 2013, vol 135, pp IX-X.

Preface

Transboundary animal diseases (TADs) are highly contagious or transmissible, have the potential to spread very rapidly irrespective of national borders, and cause high morbidity and mortality in susceptible animal populations. It has become clear that TADs seriously threaten world food security and can severely affect national economies. Therefore, the development and availability of effective vaccines and diagnostic tools is essential to control TADs.

In 2002, a meeting titled 'Vaccines for OIE List A and Emerging Animal Diseases' was organized by Iowa State University and held in Ames, Iowa. The following year a book with the same title was published by Karger. Then in 2005, a complementary meeting titled 'Marker Vaccines and Differential Diagnostic Tests in Disease Control and Eradication' was held in Ames.

Since these earlier meetings a number of advances have been made in the development of vaccines and diagnostic tools to control important TADs. An interdisciplinary group of approximately 180 scientists from academia, industry and government met from September 17-19, 2012 in Ames, Iowa for a workshop entitled 'Vaccines and Diagnostics for Transboundary Animal Diseases'. The workshop was sponsored by the US Department of Homeland Security, Science and Technology Directorate, as well as the Center of Excellence for Emerging and Zoonotic Animal Diseases (CEEZAD; www.ceezad.org/) at Kansas State University, and three other organizations—the Center for Food Security and Public Health at Iowa State University (www.cfsph.iastate.edu), the World Organization for Animal Health (OIE; www.oie.int), and the International Alliance for Biological Standardization (IABS; www.iabs.org).

The primary objective of the recent workshop was to discuss 'State of the Art' measures that the Department of Homeland Security (and it's Centers of Excellence) and the USDA can take to better position themselves to serve US and international needs related to vaccine and diagnostic tool development for significant transboundary diseases. The workshop goals included sharing progress on cutting-edge research to help inform the decision making process; presenting academic scientists with tools to help them work on translational research; enabling government officials from each agency working in this area to convey their roles and responsibilities to a broad audience; and bringing together scientists from academia, industry and government in order to stimulate cross-talk.

The papers in this publication address the roles and responsibilities of government agencies, the challenges that the biologics industry faces, and progress on the development of vaccines and diagnostics for 11 diseases. The diseases include: highly pathogenic avian influenza; exotic Newcastle disease; foot and mouth disease; Rift Valley fever; Nipah and Hendra virus; African swine fever; classical swine fever; heartwater; Q fever, Ebola, and Schmallenberg virus.

It is our hope that this publication will serve as a useful resource for researchers, government officials and industry scientists who are working to develop effective control tools for these important transboundary animal diseases.

J.A. Roth, J.A. Richt, I.A. Morozov

Roth JA, Richt JA, Morozov IA (eds): Vaccines and Diagnostics for Transboundary Animal Diseases. Dev Biol (Basel). Basel, Karger, 2013, vol 135, p 1.

SESSION I

Roles and responsibilities; facilitation and cooperation

Roth JA, Richt JA, Morozov IA (eds): Vaccines and Diagnostics for Transboundary Animal Diseases. Dev Biol (Basel). Basel, Karger, 2013, vol 135, pp 3-14.

The Role of the Department of Homeland Security, Science and Technology Directorate in the Development of Vaccines and Diagnostics for Transboundary Animal Diseases

M. Colby[1], *M. Coats*[2], *D. Brake*[3], *J. Fine*[3]

1 DVM, MS, Branch Chief, Agricultural Defense, Chemical and Biological Defense Division, Homeland Security Advanced Research and Projects Agency, Department of Homeland Security, Science and Technology Directorate

2 Program Manager, Office of University Programs, Department of Homeland Security, Science and Technology Directorate

3 PhD, Scientific Consultant, BioQuest Associates, LLC, Plum Island Animal Disease Center, Department of Homeland Security, Science and Technology Directorate

4 DVM, Principal Life Scientist, Life Sciences Operation, Health Solutions Business Unit, Science Applications International Corporation

Key words: homeland security, foot-and-mouth disease, vaccine, transboundary animal disease, agrodefense

Abstract: The development of countermeasures to support an effective response to Transboundary Animal Diseases (TAD) poses a challenge on a global scale and necessitates the coordinated involvement of scientists from government, industry and academia, as well as regulatory entities. The Agricultural Defense Branch under the Chemical and Biological Defense Division (CBD) of the Department of Homeland Security (DHS), Science and Technology Directorate (S&T) supports this important mission within the United States. This article provides an overview of the Agricultural Defense Branch's vaccine and diagnostic TAD project.

Introduction

The Food and Agriculture Organization (FAO) of the United Nations defines Transboundary Animal Diseases (TAD) as "those epidemic diseases which are highly contagious or transmissible and have the potential for very rapid spread, irrespective of national borders, causing serious socio-economic and possible public health consequences" [1]. As a result of the potential impact of these diseases on the critical agricultural infrastructure of the United States, and in the case of zoonotic diseases, the potential for disease events that involve a major public health component, US federal agencies prioritize the development of countermeasures (vaccines and diagnostics) for use against them.

Recent examples of TAD epizootics underscore their tremendous potential for societal disruption:

- In 1998-1999, the Nipah virus outbreak in Malaysia in swine and pig workers impacted 60% of the country's pig farms, and was contained after 35 weeks of control efforts, which involved the depopulation of over 1 million swine and a loss of US$ 120 million in exports [2].
- In 2001, as a result of the foot-and-mouth disease (FMD) outbreak in the United Kingdom, more than 6 million livestock were culled, with an estimated economic impact of US$ 10.7 to US$ 11.7 billion [3].
- In 2003, severe acute respiratory syndrome (SARS) began in a live animal market in Guangdong Province, China and caused global disruption. Although there were only 8,096 reported human cases, the disease spread to 30 countries and its economic impact was estimated at US$ 8 billion [4].
- In the US, public reaction stemming from media coverage of the 2009 H1N1 influenza outbreak (formerly referred to as "swine flu") was determined to have resulted in a loss of approximately US$ 156 million in revenue for lean hogs alone [5].

Unfortunately, these emerging disease incidents occur with some regularity and present credible threats to Europe, North America and beyond. In May 2012, the FAO announced that a FMD virus (FMDV) SAT2 strain was detected in the Gaza Strip, demonstrating the further regional progression of a strain that, earlier in the year, occurred in Libya and Egypt [6]. In August 2012, the FAO issued a similar warning when African swine fever (ASF) was first detected in the Ukraine; this disease is now considered enzootic in regions of the Russian Federation where it had caused an estimated US$ 240 million in economic losses in 2011 [7].

The development of countermeasures to support an effective response to TADs poses a challenge on a global scale, and necessitates the coordinated involvement of scientists from government, industry, and academia, as well as regulatory entities. Within the United States, this important mission is supported by the Agricultural Defense Branch of the Chemical and Biological Defense Division (CBD) in the Department of Homeland Security (DHS), Science and Technology Directorate's (S&T) Homeland Security Advanced Research Projects Agency (HSARPA). This article provides an overview of the DHS S&T Ag Defense Branch vaccine and diagnostic TAD project.

Background

Among US Federal Government entities, authority for defending agriculture against the threat of TADs is a shared, interagency objective. As established by the Homeland Security Act of 2002 (6 USC 101), the Secretary of DHS possesses overarching authority for federal emergency response, and is responsible for coordinating interagency emergency planning (6 USC 112) in order to ensure the effectiveness of emergency response activities (6 USC 134) [8]. The US Department of Agriculture (USDA) Secretary is authorized to respond to the presence of a threat to US livestock, and may declare an extraordinary emergency, allowing remedial actions necessary to mitigate the threat (7 USC 8306) [9]. The interrelationship of these authorities is addressed in Homeland Security Presidential Directive-7 (HSPD-7): Critical Infrastructure Identification, Prioritization, and Protection, as (1) the Secretary of DHS, is designated as the principal federal official to "lead, coordinate and integrate implementation" of efforts to protect critical infrastructure, and (2) the USDA, is designated as the Sector-Specific Agency for agriculture and food [10].

The Under Secretary for S&T possesses primary responsibility for (1) developing a national policy and strategic plan for coordinating Federal efforts for biological countermeasure development for use against emerging terrorist threats, (2) conducting research and development test and evaluation (RDT&E) relevant to DHS, and (3) "establishing priorities for, directing, funding and conducting national research, development, testing and evaluation and procurement of technology and systems" to enable detection, prevention, protection against and the response to terrorist attacks (6USC182). Similarly, S&T supports TAD countermeasure efforts for RDT&E, in accordance with Homeland Security Presidential Directive-9 (HSPD-9): Defense of US Agriculture and Food, requirements for interagency efforts to "accelerate and expand development of current and new countermeasures against the intentional introduction or natural occurrence of catastrophic animal diseases," and the associated requirement for USDA to establish a "National Veterinary Stockpile (NVS) containing sufficient amounts of animal vaccine, antiviral, or therapeutic products to appropriately respond to the most damaging animal diseases affecting the economy and that will be capable of deployment within 24 hours of an outbreak." Decisions regarding utilization and distribution of such countermeasures as part of outbreak-associated activities are the domain of USDA's Animal and Plant Health Inspection Service (APHIS) Veterinary Services (VS) [11].

S&T's mission, to "strengthen America's security and resiliency by providing knowledge products and innovative technology solutions for the Homeland Security Enterprise," is articulated in the Directorate's Strategic Plan [12]. As part of this Strategy, S&T articulates strategic goals to further define its operational priorities.

Elements of these goals, rapid delivery of innovative solutions, acquisition of needed technologies, strengthening of first responder capabilities to respond to disasters, catalyzing scientific discoveries and inventions relevant to homeland security challenge, encompass the TAD countermeasure development mission space. Within S&T, CBD focuses on protecting the Nation's infrastructure from chemical, biological, and agricultural threats and disasters. The Agricultural Defense Branch executes the CBD mission, by enhancing current capabilities and developing state-of-the-art countermeasures for high-priority TADs. This includes research and development for vaccines and diagnostics, in coordination with internal and external stakeholders.

In 2004-2005, DHS S&T and USDA first defined a joint strategy for TAD research programs, with an initial focus on foot-and-mouth disease (FMD). This activity defined the pathway for interagency collaboration in RDT&E, establishing a countermeasure candidate pipeline where promising candidates, emerging from USDA Agricultural Research Service (ARS) or academia, are selected by S&T for transition to targeted advanced development. These candidates then proceed through development activities associated with regulatory and manufacturing requirements, until such time that they are mature enough for transition to industry for full development and licensing, and are then available for acquisition by the USDA National Veterinary Stockpile (NVS). This strategic approach accounts for agency authorities, and builds upon their respective roles and strengths.

Priority Setting

HSPD-9 calls for the Secretary of DHS to lead the effort to "accelerate and expand development of current and new countermeasures against the intentional introduction or natural occurrence of catastrophic animal, plant and zoonotic diseases." These efforts, and their prioritization, are to be conducted in coordination with USDA and other interagency partners in consultation with the White House Office of Science and Technology Policy (OSTP). A component of OSTP, the National Science and Technology Council (NSTC), coordinates science and technology policy across the federal research and development enterprise. The Under Secretary for S&T serves as a co-chair for the NSTC Committee on Homeland and National Security (CHNS). This committee advises and assists the NSTC to increase the effectiveness and productivity of Federal R&D efforts in science and technology, as related to homeland and National security. The CHNS has several functions: facilitates planning, coordination, and communication among Federal departments and agencies involved in homeland or National security R&D; helps identify, define and advise the NSTC on Federal priorities and plans for homeland or National security R&D, and recommends policy options for Federal priorities; reviews and advises on Federal policy and programs that affect international efforts related to homeland or National security R&D; addresses technical programmatic and operational issues; identifies and recommends Federal priorities in National security, intelligence, and homeland security R&D, and develops options for Federal R&D budget crosscuts; and coordinates with other NSTC committees.

Since 2005, S&T has co-chaired the NSTC CHNS Foreign Animal Disease Threats Interagency Working Group (FADT IWG)[1] with USDA's ARS. Other members of the subcommittee include USDA APHIS, the Department of Health and Human Services (HHS), the Department of Defense (DoD), the Environmental Protection Agency (EPA), the National Science Foundation (NSF), the Department of the Interior (DOI), and the Smithsonian Institute. The FADT IWG coordinates interagency research programs in veterinary countermeasures; modeling; decontamination, disposal and depopulation; and basic research. In 2007, the FADT IWG established a Veterinary Countermeasures Sub-working Group to coordinate interagency research and development related to countermeasures for the National Veterinary Stockpile (NVS) in coordination with the National Veterinary Stockpile Strategic Steering Committee (NVSSSC). Specific functions of the working group include: identifying research needed to advance the development of priority countermeasures for surveillance, diagnosis, vaccination, and treatment; prioritizing the research required to develop new or improved targeted tools for each of the priority diseases identified for the NVS; reviewing new technologies and assessing their value for the future development of countermeasures to control and eradicate priority TADs; identifying government agencies to implement research projects, including strategic research collaborations and critical linkages with public, not for profit and private research organizations; identifying suitable biopharmaceutical industry partners within and outside the US; and defining and recommending short- and medium-term research plans, including milestones, deliverables and estimated costs to develop countermeasures that are fit for purpose.

The mission of developing countermeasures for use against TADs involves coordination of programmatic efforts among various elements of S&T. As described previously, the Agricultural Defense Branch, a component of CBD, is primarily responsible for funding and coordinating TAD countermeasure development within S&T. CBD is a component of the DHS Homeland Security Advanced Research Projects Agency (HSARPA), which supports homeland security research to promote changes in technologies; advance development, testing and evaluation, and deployment of technologies; and accelerate prototyping of technologies to address critical vulnerabilities. To accomplish this, HSARPA manages programs for developing homeland security technology, often partnering with elements of the DHS Office of Research and Development Partnerships. In addition to external investments, the Agricultural Defense Branch funds research within the S&T Office of National Laboratories Plum Island Animal Disease Center (PIADC) and through the Office of University Programs Homeland Security Centers of Excellence

[1] The FADT was formed in 2005 as a Subcommittee of the CHNS. In 2011, the CHNS, formed the Biological Defense Research and Development Subcommittee (BDRD) to coordinate interagency defensive research, development, testing, and evaluation (RDT&E) addressing biological threats to national security. At that time, the FADT became a working group under the BDRD.

(COEs). Scientists at PIADC support a number of TAD vaccine and diagnostic R&D programs. The DHS COEs -- the National Center for Foreign Animal and Zoonotic Disease Defense (FAZD Center), led by Texas A&M University, and the Center of Excellence for Emerging and Zoonotic Animal Diseases (CEEZAD), led by Kansas State University -- leverage an extensive network of academic and industry experts in conducting novel research and education activities that develop potential technologies for DHS and partners to target for advanced development.

Program Management

Within S&T, an Integrated Product Team (IPT) process is used to guide investments and ensure that funded projects develop capabilities that meet operational needs; analyze gaps in strategic needs and capabilities; determine operational requirements; develop programs and projects to close capability gaps and expand mission competencies. The Agricultural Defense Branch, in conjunction with the DHS Offices of Health Affairs and Infrastructure Protection, chairs the Food and Agriculture Sub-IPT. This Sub-IPT includes members from within DHS Components (including Customs and Border Protection) as well as members from other Federal agencies whose mission includes agricultural defense. The mission of the Food and Agriculture Sub-IPT is to prioritize capability gaps and recommend S&T investments that increase preparedness against agricultural threats through similar measures. Topics considered within the Food and Agriculture Defense mission space include threats to animal health, the use of food as a vehicle for chemical and biological threats, and the protection of the food and agriculture critical infrastructures.

Under the Sub-IPT, Project IPTs are established to integrate input, analysis, and decision making from a variety of perspectives and constituencies; in particular, customers, stakeholders, and users who are vital to the successful transition of products and knowledge from S&T to the ultimate end users. To support TAD countermeasure development, there are two Project IPTs subordinate to the Food and Agricultural Defense Sub-IPT: the Foreign Animal Disease (FAD) Vaccines and Diagnostics Project IPT and the Agricultural Screening Tools Project IPT. These Project IPTs provide a mechanism to review critical research gaps identified by stakeholders, the progress made by (or known to) all participating agencies to date, and to develop a plan to address the highest priority gaps through interagency partnerships. Further, the Project IPTs provide an enduring capability for TAD countermeasure advanced development by employing a gated decision-making process with clearly defined transition endpoints for individual products.

The FAD Vaccines and Diagnostics Project IPT addresses the need for the development of effective veterinary countermeasures against high-priority TADs, in partnership with USDA and industry. It serves as a coordinating body to steer S&T strategic investments in RDT&E targeted to address these capability gaps. The projects administered by this IPT fall within three broad categories: FMD vaccines, countermeasures for other TADs, and companion diagnostics for each vaccine allowing the differentiation of infected from vaccinated animals (DIVA). RDT&E of countermeasures for FMD are the highest priority for S&T, given the highly infectious nature of the virus and its potential to infect multiple livestock species of economic importance in the United States. In addition, the DHS owned and operated PIADC is the only facility within the US that is capable of conducting research with FMD. Research in this area began in 2005, with near term efforts focused on studies to perform enhanced characterization of the efficacy of vaccine antigen (housed within the North American FMD Vaccine Bank), and similar studies involving commercial FMD vaccines. Mid-term efforts involved enhanced characterization and the issuance of import permit licenses for sale and distribution for commercial off-the-shelf vaccines and diagnostics produced outside of the US Long-term efforts include the development of new serotype and subtype specific, marked, molecular vaccines that do not require the use of live FMD virus. S&T maintains a robust pipeline of these vaccine candidates and is aggressively pursuing a development schedule designed to advance them through an established late-stage development and USDA regulatory pathway.

In addition to FMD, the FAD Vaccines and Diagnostics Project IPT maintains programs intended to develop countermeasures for other high-consequence TADs. Targets are prioritized pathogens, as identified by USDA APHIS and the NVSSSC, and through interagency working groups, e.g. the FADT. This Project IPT portfolio includes near-, mid- and long-term projects for swine (classical swine fever and African swine fever) and zoonotic (Rift Valley fever and Henipavirus) targets.

Lastly, as part of the FAD Vaccines and Diagnostics Project IPT, S&T is pursuing a parallel diagnostic development program to ensure that complementary diagnostic technologies are available to maximize the value of the DIVA-friendly aspects of the molecular FMD vaccines. These efforts are intended to minimize the potential for trade restrictions associated with the use of these novel vaccines. Efforts also include the development of next-generation, high-throughput serology and molecular-based technologies for use in surge situations by the National Animal Health Laboratory Network (NAHLN).

In a manner similar to that of the FAD Vaccines and Diagnostics Project IPT, the Agricultural Screening Tools (AST) Project IPT guides the development of standardized protocols and technologies for use by the National Animal Health Laboratory Network (NAHLN) and the livestock industry. The objective of the AST Project is to provide rapid detection and field identification of high-priority pathogens that threaten US livestock. In addition to laboratories, these tests may be used at border labs and inspection houses, animal collection points, and other environments such as farms, by multiple government agencies to detect high-priority TADs during outbreaks affecting the US agriculture critical infrastructure. This project leverages interagency translational discovery research and diagnostic development conducted by USDA ARS and APHIS. Placement of diagnostic testing capabilities closer to sites of animal production and importation (or entry into the country) will enhance the National capability to detect, respond to and recover from an outbreak of TADs in the agriculture sector.

First successes

As part of the objectives of the FAD Vaccines and Diagnostics Project IPT, S&T successfully facilitated access to FMD vaccines manufactured overseas through a relationship with Transboundary Animal Biologics, Inc. (TABI), a not-for-profit company based in Ames, Iowa. Established in 2008, TABI was funded by S&T to secure USDA licenses/permits for TAD vaccines and diagnostics. Through this mechanism, in July 2011, an import permit was granted for Bioaftogen®, a quadravalent FMD vaccine manufactured by Biogénesis - Bagó, a veterinary vaccine manufacturer in Argentina. This vaccine covers FMD serotypes often found in South America (A24 Cruzeiro, A2001 Argentina, C3 Indaial, and O1 Campos) and is to be used only under the direction of USDA.

As described earlier, S&T has invested in RDT&E efforts to advance the application of novel vaccine technologies for use as FMD countermeasures. In May 2012, USDA issued a conditional license for FMD Virus Vaccine, Serotype A24, Live Adenovirus Vector (AdA24) a molecular vaccine developed by S&T through a public-private partnership that also included USDA and industry. As opposed to traditional inactivated vaccines, this molecular product is derived from a replication deficient viral vector, offers rapid, serotype-specific protection, is DIVA (differentiates in infected from vaccinated animals) compatible, only contains a piece of the FMD viral genome, and may be produced domestically. This achievement resulted from collaboration with the vaccine industry and veterinary vaccine regulatory authority (USDA Center for Veterinary Biologics) and is the culmination of over ten years of effort. It is recognized as the first successful FMD vaccine technology developed into a commercial product in over fifty years.

Ongoing and future activities

S&T collaborates with USDA and extramural partners, including DHS COE, National Labs, and industry, to identify promising candidates and leverage complementary assets. Through these partnerships, S&T is able to continuously draw from exploratory technology investment products for advanced development. For example, based on the initial success of the AdA24 FMD molecular vaccine, the PIADC Targeted Advanced Development Branch, working collaboratively with USDA and an industry partner, has built a pipeline of AdFMD vaccine candidates for other high-priority FMDV serotypes/subtypes. Candidates that pass an efficacy threshold are transitioned to the master seed phase for subsequent development studies associated with regulatory licensing requirements. Molecular re-engineering of candidates that fail initial efficacy testing is being conducted to improve vaccine potency. In parallel, AdFMD manufacturing process improvement studies are being performed to reduce vaccine cost of goods. Investments in and initial testing of additional promising molecular-based vaccine platforms for FMD, such as recombinant, subunit antigen production (baculovirus, E. coli, tobacco plants) and DNA vaccines, are ongoing. For high-priority swine TADs, S&T has invested in USDA ARS rationally designed live, attenuated virus approaches that apply reverse genetics to identify vaccine candidates for classical swine fever (CSF) and ASF. Additional ASF collaborative projects with the FAZD COE and Texas A&M University are focused on the identification of lead vaccine candidates based on reverse vaccinology, bioinformatics technology, and the development of a multi-component vaccine. For zoonotic TADs, S&T is funding Rift Valley fever vaccine and diagnostic efforts with other government agencies (USDA, DoD), and has identified two lead vaccine candidates for Henipavirus that may guide future strategic partnerships with industry.

Working with the Department of Homeland Security

S&T issues open, competitive solicitations seeking expertise, within the US and globally, to find solutions for addressing the most pressing challenges in homeland security. Partnerships are sought to address capability gaps in a number of areas, including TAD countermeasure development. DHS utilizes a variety of funding strategies to administer awards, selecting the method of partnership based on the stage and progress of the research or technology in development. There are several vehicles through which interested parties may work with S&T: contracts, grants and cooperative agreements, and cooperative research and development agreements (CRADAs).

A Federal contract is defined as an agreement between the Federal Government and any person for the exchange of real or personal property or services. Contracts are utilized as an enabler of the systems acquisition management environment, which includes structuring programs to account for research and development, test and evaluation, production, deployment, operations support, and disposal. In doing so, DHS determines the type of contract most appropriate to achieve the desired capability or development goal. S&T executes contracts to perform mission critical research and development activities:

- The Long Range Broad Agency Announcement (LRBAA) is used to seek R&D projects for revolutionary, evolving, and maturing technologies that demonstrate the potential for significant improvement in homeland security missions and operations. As part of the 2012 LRBAA, the Agricultural Defense Branch solicited proposals for new, cost-effective, biological-based countermeasures for TAD pathogens affecting major domestic livestock species. Specific areas of interest are cattle and swine product candidates based on molecular vaccine platforms previously shown to be safe and effective against other infectious or human biodefense disease targets, and novel biological-derived agents with an established immune-based mechanism of action.

- DHS issues competitive solicitations through the Federal Business Opportunities (Fed Biz Ops) website (https://www.fbo.gov/).

Federal grants and cooperative agreements are financial assistance awards that aid recipient(s) in carrying out activities of public purpose. DHS utilizes grants and cooperative agreements to support preparation, prevention, and response activities of state, local, and tribal communities and to support research and education focused on addressing homeland security challenges. As the research arm of the Department, S&T utilizes research and education grants and cooperative agreements to encourage innovation and facilitate early-stage technology development. The Federal Grant and Cooperative Agreement Act (P.L. 97-258) (31 USC Sec. 6301.-6305.) [13] guides agencies to use a grant agreement as the legal instrument reflecting a relationship between the United States Government and a state, local government, or other recipient when: (1) the principal purpose of the relationship is to transfer a thing of value to the state or local government or other recipient to carry out a public purpose of support or stimulation authorized by a law of the United States instead of acquiring (by purchase, lease, or barter) property or services for the direct benefit or use of the United States Government; and (2) substantial involvement is not expected between the executive agency and the state, local government, or other recipient when carrying out the activity contemplated in the agreement. DHS will use a cooperative agreement when substantial involvement is expected in the implementation of the awards activities. Various websites offer grant opportunities:

- Primary website that assists applicants in finding and applying for federal grants: www.grants.gov:
- Homeland Security COEs solicit proposals for competitively funded research and development and education projects. These projects relate to transformative, evolving and/or maturing R&D technologies that demonstrate significant improvement in the Center's mission or operations of Center stakeholders. Proposals involving TAD countermeasure development fall within the domain of the FAZD Center and CEEZAD COEs.
  - COEs' competitive requests for proposals: http://fazd.tamu.edu/ and http://www.ceezad.org/ and www.hsuniversityprograms.org
  - Career Development Grants (CDGs) are competitively awarded to accredited universities in the US to provide financial support to students interested in HS-STEM (Homeland Security Science, Technology, Engineering, and Math)-related research and careers: www.grants.gov
  - Summer Internships provided to rising juniors and seniors for up to 10 weeks during the summer to conduct research in DHS mission-relevant research areas: http://www.orau.gov/dhsinternships/
  - Minority Serving Institutions (MSI) Grants and Awards encourage partnerships with S&T COEs: www.hsuniversityprograms.org
  - MSI Summer Research Team Project grants fund research opportunities at DHS COEs that increase and enhance the scientific leadership at MSIs in DHS-relevant research areas: http://www.orau.gov/dhsfaculty/

- Scientific Leadership Awards (SLA) fund the development of HS-STEM teaching initiatives, curriculums, and scholarships at MSIs: www.grants.gov

- S&T International Cooperative Programs Office Grants are designed to augment and complement, through international research and collaboration, the depth and breadth of homeland security science and technology research: www.grants.gov

CRADAs are a key feature of technology development management. Under a CRADA, Federal and non-Federal entities are encouraged to leverage research and development budgets to advance critical technologies through cost sharing practices. CRADAs are commonly used for technology transfer between the US government and industry, and may be used in any aspect of a product life cycle where RDT&E activities occur. CRADAs are authorized by the Federal Technology Transfer Act of 1986, as amended and codified by 15 USC 3710a [14]. DHS, as an executive agency under 5 USC 105, is a Federal agency for the purposes of 15 USC 3710a and may enter into a CRADA. [14, 15] DHS delegated the authority to conduct CRADAs to S&T and its laboratories. Through this mechanism, Federal parties may provide personnel, services, facilities, equipment, intellectual property, or other resources with or without reimbursement. The non-Federal parties may provide funds, personnel, services, facilities, equipment, intellectual property, or other resources toward the conduct of specified research and development efforts that are consistent with the missions of the component or laboratory.

Conclusions

At the 2011 annual meeting of the United States Animal Health Association (USAHA), Dr. John Clifford, Deputy Administrator for USDA APHIS Veterinary Services, announced that APHIS had shifted its thinking with regard to FMD control measures in the US He explained that the traditional response of stamping out may no longer be appropriate, adding that vaccination was currently under consideration, among several tools, as part of a FMD control strategy. These sentiments are echoed in the USDA APHIS National Center for Animal Health Emergency Management (NCAHEM) Foreign Animal Disease Preparedness & Response Plan (FAD PReP), where the goals of a US response to a domestic incursion of FMD are specified as "to (1) detect, control, and contain FMD in animals as quickly as possible; (2) eradicate FMD using strategies that seek to stabilize animal agriculture, the food supply, the economy, and protect public health; and (3) provide science- and risk-based approaches and systems to facilitate continuity of business for non-infected animals and non-contaminated animal products" [16]. The authors of the plan state that "the use of emergency vaccination strategies may be considered in an FMD outbreak." Dr. Clifford's announcement and the technical successes of the program validate the TAD countermeasures development approach executed by S&T. For FMD and other high-consequence TADs, it is clear that the breadth of the program is warranted in order to ensure the availability of countermeasures for emergency response, as part of the USDA APHIS paradigm, should there be a need for them.

References

1 EMPRES TADS. Rome: Food and Agriculture Organization of the United Nations. Retrieved September 18, 2012. Available from: http://www.fao.org/ag/againfo/programmes/en/empres/diseases.asp.

2 Daszak P, Cunningham AA, Hyatt AD: Anthropogenic environmental change and the emergence of infectious diseases in wildlife. Acta Tropica 2001;78(2): 103-116.

3 Thompson D, Muriel P, Russell D, Osborne P, Bromley A, Rowland M, et al: Economic Costs of the Foot and Mouth Disease Outbreak in the United Kingdom in 2001. Rev Sci Tech OIE 2002; 21(3): 675-687.

4 Institute of Medicine (US): Addressing Foodborne Threats to Health: Policies, Practices and Global Coordination. Workshop Summary. Washington: The National Academies Press; 2006.

5 Attavanich W, McCarl BA, Bessler D: The Effect of H1N1 (Swine Flu) Media Coverage on Agricultural Commodity Markets. Applied Economic Perspectives and Policy 2011 33(2): 241-259.

6 Foot and mouth disease reported in Gaza Strip. Rome: Food and Agriculture Organization of the United Nations. Retrieved September 18, 2012. Available from: http://www.fao.org/news/story/en/item/141694/icode/.

7 Pig disease outbreak in Ukraine a major warning sign. Rome: Food and Agriculture Organization of the United Nations. Retrieved September 18, 2012. Available from: http://www.fao.org/news/story/en/item/154778/icode/.

8 "Homeland Security Act of 2002," Title 6 US Code, Pts. 101. 2002 ed.

9 "Animal Health Protection Act," Title 7 US Code, Pts. 8306. 2002 ed.

10 The White House, Homeland Security Presidential Directive 7 (HSPD-7): Critical Infrastructure Identification, Prioritization, and Protection (Dec. 17, 2003).

11 The White House, Homeland Security Presidential Directive 9 (HSPD-9): Defense of United States Agriculture and Food (January 30, 2004).

12 Department of Homeland Security: DHS Science & Technology Directorate - Strategic Plan. Washington, D.C.: Department of Homeland Security, 2011. http://www.dhs.gov/xlibrary/assets/st/st-strategic-plan.pdf (accessed September 19, 2012).

13 31 USC Chapter 63 – Using Procurement Contracts and Grants and Cooperative Agreements

14 "Utilization of Federal Technology," Title 15 US Code, Pts. 3710a. 2011 ed.

15 "Executive Agency," Title 5 US Code, Pts. 105. 2011 ed.

16 United States Department of Agriculture. Foreign Animal Disease Preparedness & Response Plan. Washington, D.C.: United States Department of Agriculture, June 2012 (draft). http://www.aphis.usda.gov/animal_health/emergency_management/downloads/fmd_responseplan.pdf (accessed September 25, 2012).

Dr. Michelle Colby, Mail Stop #0201, Chem/Bio Division, 245 Murray Lane SW, Washington, DC 20528
Fax: 202-254-6164
Email: Michelle.Colby@hq.dhs.gov

Roth JA, Richt JA, Morozov IA (eds): Vaccines and Diagnostics for Transboundary Animal Diseases. Dev Biol (Basel). Basel, Karger, 2013, vol 135, pp 15-22.

The Perspective of USDA APHIS Veterinary Services Emergency Management and Diagnostics in Preparing and Responding to Foreign Animal Diseases – Plans, Strategies, and Countermeasures

J.R. Díez, D.K. Styles

Emergency Management and Diagnostics, Veterinary Services, Animal and Plant Health Inspection Service, United State Department of Agriculture, Riverdale, MD, USA

Key words: HPAI, H5N1, avian influenza, control strategies

Abstract: The United States Department of Agriculture (USDA) Animal and Plant Health Inspection Service (APHIS) Veterinary Services (VS) is charged with monitoring, controlling, and responding to select reportable diseases and all foreign animal diseases. Emergency Management and Diagnostics (EM&D) oversees Foreign Animal Disease (FAD) preparedness and response. In order to effectively prepare for and respond to FADs, such as highly pathogenic avian influenza and foot-and-mouth disease, VS develops plans, strategies, and policies to effectively combat an intrusion. USDA APHIS VS has made significant gains in preparedness and response planning. However, much remains to be done especially in surveillance, diagnostic tools, and vaccines. There are significant needs for novel medical technologies to improve diagnostic capabilities and offer additional approaches for FAD response.

Introduction

The United States Department of Agriculture (USDA) Animal Plant Health Inspection Service (APHIS) Veterinary Services (VS) division is charged with monitoring, controlling, and responding to select reportable domestic diseases and all foreign animal diseases (FADs). The division overseeing response to FADs is

Emergency Management and Diagnostics (EM&D) which includes three units responsible for a multitude of activities relating to mitigation, diagnostics, and response to FAD threats. This paper addresses how APHIS Veterinary Services prepares and responds to FADs.

Emergency Management and Diagnostics

The National Center for Animal Health Emergency Management (NCAHEM) is composed of three subunits:

- Preparedness and Incident Coordination develops response strategies; collaborates with stakeholders in developing those strategies, and provides national perspective in the coordination of FAD incidents;
- Interagency Coordination identifies Government partners and potential resources, and collaborates with interagency and international working groups;
- The National Veterinary Stockpile (NVS) stockpiles veterinary countermeasures, and provides States, Tribes, and Territories logistics support for FAD incidents;

The National Veterinary Services Laboratories (NVSL) provides confirmatory and reference laboratory diagnostic services and establishes standards for national FAD testing for laboratories belonging to the National Animal Health Laboratory Network (NAHLN). Another VS unit, The Center for Epidemiology and Animal Health's National Surveillance Unit, works with others VS units and establishes the protocols for sampling frequency; sampling size; and target populations to be sampled.

The Center for Veterinary Biologics (CVB) is responsible for the evaluation, inspection, and licensing or permitting of all veterinary biologics including vaccines and diagnostic test kits, distributed in the United States. The CVB implements the provisions of the Virus-Serum-Toxin Act to ensure that veterinary biologics available for the diagnosis, prevention, and treatment of animal diseases, including FADs, are pure, safe, potent, and effective.

Mitigation, Preparedness, and Response Activities

Veterinary Services engage in active and passive surveillance programs to detect pathogens of concern. EM&D monitors existing active surveillance programs for select FADs such as Highly Pathogenic Avian Influenza (HPAI), where testing results are reported by State approved laboratories based on submissions from industry and regulatory sources. For most FADs, EM&D relies chiefly on passive surveillance through a network of APHIS and Food Safety and Inspection Service (FSIS) veterinary medical officers (VMOs), State regulatory VMOs, and Federally accredited private sector veterinarians who report suspect cases. The initial response strategy is outlined in VS memorandum 580. 4 "Procedures for the Investigation of

Potential Foreign Animal Disease/Emerging Disease Incidents (FAD/EDI)" which details how investigations are conducted, samples submitted, and details the chain of communication that should occur between Federal and State animal health authorities, NVSL, and the NAHLN laboratories. Depending on the suspected FAD, samples may be forwarded directly to the NVSL (e.g., foot and mouth (FMD) disease), or initially analyzed at an approved NAHLN laboratory with suspect samples forwarded to NVSL for confirmatory diagnostics (e.g., HPAI).

APHIS VS conducts preparedness and response activities through the NCAHEM which develops industry specific response plans and countermeasures that rely heavily on stakeholder input. During an outbreak, NCAHEM provides the guidance and coordination for responding to an outbreak in compliance with Federal regulation; coordinates the event on a National level; assists States in their response and control efforts; and helps to deploy qualified Federal personnel to assist in the mission. NCAHEM, in conjunction with local officials, determines what strategies will be undertaken, and how and when to deploy countermeasures such as vaccines. All animal health emergency responses are conducted using the Incident Command System (ICS) model, with a Unified Field Command (Federal/State) established locally to oversee the response. The ICS is a standardized tool for meeting the demands of an emergency situation. APHIS would also reach out to other Federal agencies for support (e. g., Department of Homeland Security; Environmental Protection Agency, etc.) in accordance with the National Response Framework (NRF).

The NVS is a national repository of veterinary supplies that can be deployed within 24 hours to augment State, Tribe, and Territory resources during a response to a damaging animal disease outbreak. The stockpile contains critical countermeasures, such as personal protective equipment (PPE) and decontamination materials; animal handling and depopulation equipment; animal vaccines; antiviral medications for agricultural responders; and response support services, including depopulation, disposal, and decontamination (3D) efforts.

The NVS became operational in 2006 and the program continues to research and add new technologies, including animal vaccines. Since 2003, several US interagency working groups have evaluated animal diseases and ranked their danger. Based on these early efforts, in 2005 the NVS' Intra-governmental Strategic Steering Committee identified and ranked the most dangerous diseases for which the NVS needed to have countermeasures. The group used the following criteria to select and rank the diseases:

- Epidemic Potential - Ability to shed, spread, and rapidly infect target species;
- Economic Impact - Loss of revenue to a region, one or more agricultural segments (e.g., beef, dairy, broilers) and associated industries;
- Trade Impact - Loss of revenues due to trade restrictions imposed by one or more trade partners;
- Zoonotic Potential - Ability of an animal disease to spread and cause morbidity and/or mortality to a small or large number of people;

- Morbidity and Mortality - The virulence potential of a pathogen and its ability to cause sub-clinical disease, moderate disease, severe disease, and/or mortality;
- Cross-species Potential – The ability of a pathogen to cross a species barrier, infect, and cause disease in other animal species, including establishing a reservoir in important domestic or wildlife species;
- Inability to Detect Rapidly - Availability of very specific and sensitive tests to rapidly detect the pathogen in the field;
- Inability to Vaccinate - Availability of vaccines that have the ability to control and eradicate the pathogen (i.e., VS would be unable to implement a vaccine strategy if there are no vaccines, if available vaccines are not "marked", or do not prevent shed and spread, colonization of target tissues, carrier animals).

The group defined the following as the top five most dangerous animal disease threats:

1. Highly pathogenic avian influenza
2. Foot-and-Mouth disease
3. Rift Valley Fever
4. Exotic Newcastle disease
5. Nipah and Hendra viruses

The list will serve as guidance for countermeasures acquisitions, including animal vaccines.

Discussion

APHIS VS EM&D has made significant progress in developing novel response technologies against diseases of concern but significant challenges remain. However, adequate resources are needed to address those challenges. Often times, gains are made when a particular issue reaches some level of notoriety. For example, the real and perceived potential threat to public health posed by highly pathogenic avian influenza subtype H5N1 (HPAI H5N1) resulted in significant monetary resources being made available to develop novel response tools and diagnostics in poultry. These tools include highly sensitive and specific field diagnostic tests (e. g., adaptation of available avian influenza assays to high throughput testing platforms - implementation of the semi-automated processes markedly increased the testing capacity of the NAHLN), and mass depopulation technology (e. g., water-based foam depopulation technology) that quickly depopulates poultry and is relatively safe for human operators. Consequently, APHIS is now better prepared to quickly address more FADs that may threaten poultry health. Conversely, advances in diagnostic and response technologies for FADs affecting hoofstock have not been as fortunate. This is especially true for catastrophic and highly infectious diseases such as FMD. There is an urgent need to develop novel diagnostic platforms for FMD and an equal or greater need for sufficient vaccine stocks and mass depopulation and disposal technologies [1].

APHIS and the Department of Homeland Security have jointly funded many projects for depopulation (e. g., mobile center-channel pneumatic captive bolts for cattle; on-farm CO2 depopulation for swine) and disposal (gasification of carcasses technology), that will produce adequate technologies and research data. These projects will help to provide the necessary tools for responders to address current and future outbreaks. Improvements in FMD vaccine technology have been slow, with the exception of the development of a single topotype human adenovirus-5 vectored FMD vaccine [2]. Thus, agriculture is globally dependent on inactivated FMD vaccine products for the foreseeable future [3,4].

FMD is arguably the most economically devastating disease of livestock [5], and the United States has historically committed to using only a "stamping-out" approach to eradicate the disease. Stamping out was an appropriate action to take with the livestock numbers and systems that were in place when the US had its last outbreak of FMD in 1929. In the intervening years, the US population of susceptible livestock has markedly increased, and farming methods have shifted from low density stocking rates to large concentrated animal feeding operations (hoofstock population is approximately 174 million, USDA NASS Data 2007) [6,7]. Moreover, vast numbers of livestock move rapidly across state and international borders compared to the past decades [8]. It is estimated that 625,000 swines are in transit daily within the United States [National Pork Board, unpublished data].

The use of only a "stamping out" policy during a disease outbreak presents significant challenges due to the need to rapidly and humanely depopulate and dispose of large numbers of animals [9,10]. With the current technologies it is physically and environmentally impractical to rapidly dispose of large numbers of animal carcasses [11-13]. In the 2010-2011 FMD outbreak in the Republic of Korea, animal health officials reportedly faced depopulation challenges [10] and environmental problems involving the leakage of leachate into the water table due to improperly lined burial pits [14,15]. Finally, large scale depopulation efforts may severely cripple or even destroy an industry. By some estimates, the swine industry in the Republic of Korea may take decades to recover.

Given the magnitude of these challenges, it was clear to VS, State Animal Health Officials, and industry leaders that a "stamping out only" approach was impractical and unrealistic for many diseases where animal sparing countermeasures (i. e., vaccine) could be employed. Realistically, any response will need to be tailored to the size of the outbreak and the circumstances surrounding it, and stamping out will be part of an escalating and intensifying response. But the magnitude of the tasks of depopulation and disposal when the outbreak is of an immense scale calls for an alternate strategy to stamping out alone.

Veterinary Services will use FMD vaccine as a tool to control an outbreak with four basic strategies for response

- **Stamping out with no vaccination** – This approach is really only useful for a highly circumscribed and limited outbreak involving a relatively small number of animals.

- **Stamping out with emergency vaccination to slaughter/kill** – In this approach, vaccine is applied in animals that either proceed to slaughter or are depopulated and processed via hygienic disposal, and it shortens the time for eligibility to export.
- **Stamping out with emergency vaccination to live** – This strategy preserves the integrity of the food supply chain and provides regulators a longer timespan to effect an eradication; the disease is halted and animals complete their productive lifespan; vaccination eventually ceases, but this approach closes international export markets for an indeterminate amount of time.
- **Vaccination to live with no stamping out** – Vaccination would become a routine production procedure and the time to eradication (if ever) would be indefinite; new export agreements would need to be developed, but it would severely limit the production of key industries such as swine which rely heavily on export.

Due to the complexity of the FMD virus (which consists of 7 different serotypes and multiple topotypes), it is challenging for APHIS to anticipate which topotypes to stock in quantities sufficient to address the needs of any outbreak. This is one of the main limitations of the current inactivated vaccine technology. Based on a risk analysis for North America, the Technical Committee of the North American Foot-and-Mouth Vaccine Bank (Bank) recommends which topotypes to stock, in the form of vaccine antigen concentrate (VAC) used to produce emergency vaccines of high potency. The Bank's supplies are based on the old model of very selective and restricted use of vaccine. Therefore, stocks are far below what would be required to address a livestock dense state or multi-state outbreak. We are hopeful that commercial FMD vaccines that have a somewhat reduced potency would help to bridge the gap created by the limited supply of emergency vaccines until more technologies are perfected [16].

To add commercial FMD contracts to the inventory, the NVS initially engages the CVB to determine whether a prospective foreign manufacturer meets requirements for vaccine importation. The manufacturer must demonstrate to the CVB that its product is pure, safe, potent, and effective to be eligible for permitting. CVB personnel reviews the facility's manufacturing practices, quality assurance, and other data supplied to meet permitting requirements. Once permitted, the NVS can engage the commercial vaccine manufacturer in a variety of contract arrangements.

Conclusion

This paper has shown the activities of APHIS VS EM&D in preparing for and responding to FADs, in particular FMD in the United States. A great deal of work remains to be done to ensure that plans, strategies, and countermeasures being developed match the requirements of an actual outbreak. Additional capabilities are necessary to bolster the preparedness of the United States. More resources are needed to develop better diagnostic platforms and response technologies. Novel vaccine technologies that are nimble, flexible, and do not require foreign production facilities are needed. The vectored products offer some promise. However, immunological performance with different topotypes, duration of immunity, and boosting of immunity needs further evaluation. Novel and highly versatile technologies such as DNA vaccines may be good alternatives to inactivated vaccines, but more intensive research is needed to drive innovation in the marketplace to be better prepared for what will inevitably occur in the future. An ideal vaccine would have the following characteristics: be domestically manufactured; relatively inexpensive (≤$1. 00/dose); have the flexibility to rapidly change (≤72 hours) the target epitope(s); demonstrate a rapid onset of protection (≤72 hours) with a single dose; require only yearly (or longer) boostering; be effective in a variety of livestock species, ages, and breed types; prevent infection in addition to reducing clinical signs; and provide a broad umbrella of immunity across closely related but poorly cross-reactive topotypes; and provide the ability to differentiate infected from vaccinated animals.

References

1. Torres A, David MJ, Bowman OP: Risk management of international trade: emergency preparedness. Rev sci tech Off Int Epiz 2002;21(3):493-498.
2. Grubman MJ, Moraes MP, Schutta C, Barrera, Neilan J, Ettyreddy D, et al: Adenovirus serotype 5-vectored foot-and-mouth disease subunit vaccines: the first decade. Future Virol 2010;5(1):51-64.
3. Doel TR: FMD vaccines. Virus Res 2003;91:81-99.
4. Barnett PV, Carabin H: A review of emergency Foot-and-mouth disease (FMD) vaccines. Vaccine 2002;20:1505-1514.
5. Paarlberg PL, Seitzinger AH, Lee JG, Mathews KH: Economic Impacts of Foreign Animal Disease. USDA Economic Research Service 2008. Report No. 57.
6. Fraser D: Farm animal production: changing agriculture in a changing culture. J Appl Ani Welfare Sci 2001;4(3):175-190.
7. Elbakidze L, Highfield L, Ward M, McCarl BA, Norby B: Economic analysis of mitigation strategies for FMD introduction in highly concentrated animal feeding regions. Rev Econ 2009;31(4):931–950.
8. Bajardi P, Barrat A, Savini L, Colizza V: Optimizing surveillance for livestock disease spreading through animal movements. J R Soc Interface 2012;9(76):2814-2825.
9. Styles DK, Gackstetter GD: Mass Animal Fatalities in Response to Animal Health Emergencies: Depopulation and Disposal of Livestock. in Death in Large Numbers. 2012. American Medical Association. Chapter 7: 201-218.
10. Live burial of pigs provokes mass uproar. The Korea Times National, January 16, 2011.
11. Swine Carcass Disposal Options for Routine and Catastrophic Mortality. (Issue Paper) Council for Agricultural Science and Technology. No. 39. 2008.
12. Ruminant Carcass Disposal Options for Routine and Catastrophic Mortality. (Issue Paper) Council for Agricultural Science and Technology. No. 41. 2009.
13. Gwyther CL, Williams AP, Golyshin PN, Edwards-Jones G, Jones DL: The environmental and biosecurity characteristics of livestock carcass disposal methods: A review. Waste Manage 2011;31(4):767-778.
14. Park S: Pigs burial raises concerns over water contamination. The Korea Times National, January 3, 2011.

15. Kim H, Choi J, Kim M, Choi J, Lee M, Lee H, Jeon S, Bang S, et al: Development of Chemical Indicators of Groundwater Contamination Near the Carcass Burial Site. American Geophysical Union, Fall Meeting. 2011. #H41B-1029.
16. Pandya M, Pacheco JM, Bishop E, Kenney M, Milward F, Doel T, Golde WT: An alternate delivery system improves vaccine performance against foot-and-mouth disease virus (FMDV). Vaccine 2012;30(20):3106-3111.

José R. Díez; Darrel K. Styles; Emergency Management and Diagnostics, Veterinary Services, Animal and Plant Health Inspection Service, United State Department of Agriculture;
4700 River Road, Unit 41, Riverdale, MD 20737
Tel: 301-851-3595.
Jose.R.Diez@aphis.usda.gov
Darrel.K.Styles@aphis.usda.gov

Roth JA, Richt JA, Morozov IA (eds): Vaccines and Diagnostics for Transboundary Animal Diseases. Dev Biol (Basel). Basel, Karger, 2013, vol 135, pp 23-37.

Agricultural Research Service: Biodefense Research

C.G. Gay

United States Department of Agriculture, Agricultural Research Service, Office of National Programs, Animal Production and Protection, Beltsville, MD, USA

Key words: vaccines, diagnostics, countermeasures, one health, stockpile

Abstract: The National Animal Health Program at the Agricultural Research Service (ARS), United States Department of Agriculture (USDA), includes research programs dedicated to the defense of animal agriculture against the treat of biological agents with the potential of significant economic harm and/or public health consequences. This article provides a summary of the program and identifies its relevance to national initiatives to protect livestock and poultry as well as global food security. An introduction to setting research priorities and a selection of research accomplishments that define the scope of the biodefense research program is provided.

Introduction

Potential acts of agroterrorism constitute major threats to our Nation's agricultural resources, foods, and public health, thus presenting a significant challenge to our national security. The United States Government must plan, based on the assumption that what is likely to occur will occur, and has therefore significantly expanded its efforts to improve the Nation's ability to rapidly recognize and respond to intentional, accidental, or natural occurrences that may result in a pandemic or epizootic infectious disease. Although the United States is fortunate that very few incursions have occurred on U.S. territories, the risk of a biological agent introduction has greatly increased, mainly as a result of increased global movement of people and animals. With increasing movement of animals and agricultural products across international borders, the traditional tools used by government action and regulatory agencies worldwide to mitigate infectious disease outbreaks are becoming more and more challenged. The threats are now directly linked to the disease outbreaks that occur every day throughout the globe. The threat to our food supply and the health of citizens worldwide continue to impact global security, resulting in potential economic devastation and political instability. We have repeatedly witnessed the spread of emerging diseases such as foot-and-mouth disease, highly pathogenic H5N1 avian influenza and, most recently, pandemic H1N1 influenza virus, across the globe. To address these concerns, the Agricultural Research Service recognizes the need to maintain a comprehensive animal health biodefense research program to confront the biological threats facing our Nation.

Agricultural Research Service

Research agency

The Agricultural Research Service (ARS) is the principal in-house research agency of the United States Department of Agriculture (USDA). ARS is one of four agencies in the Research, Education, and Economics (REE) mission and is charged with extending the Nation's scientific knowledge with research projects in agriculture, human nutrition, food safety, natural resources, and the environment. ARS supports more than 2,100 scientists working on approximately 1,050 permanent research projects located at over 100 locations in the United States and five overseas laboratories.

Role and function

ARS conducts research to find solutions to problems of high national priority that impact the American people on a daily basis. ARS often implements high-risk, long-term, scientific projects to achieve significant breakthroughs in important problem areas. ARS research programs also complement the work of State Colleges and Universities, State Agricultural Experiment Stations, other Federal agencies, and the private sector. Mechanisms for addressing state and local issues are already in place; therefore, research activities at ARS focus on issues having a regional or national scope and where there is a clear federal role. ARS also provides research support to USDA action and regulatory agencies and to a number of other Federal agencies, including the Departments of Homeland Security, State, and Defense, the Centers for Disease Control and Prevention, the Food and Drug Administration, and the Environmental Protection Agency.

National Animal Health Program

Organization

The research portfolio at ARS is allocated and managed under 19 National Programs [1]. These programs serve to bring direction, coordination, and communication to 800 research projects carried out by ARS. The National Programs focus on the relevance, impact, and quality of ARS research. The National Animal Health Program is one of four National Programs under Animal Production and Protection. Although the National Animal Health Program is managed independently, many of the research projects within the program are strategically linked to other national programs to maximize the use of scientific expertise and resources. The National Animal Health Program currently has 100 scientists working at 11 locations in the United States (Fig. 1).

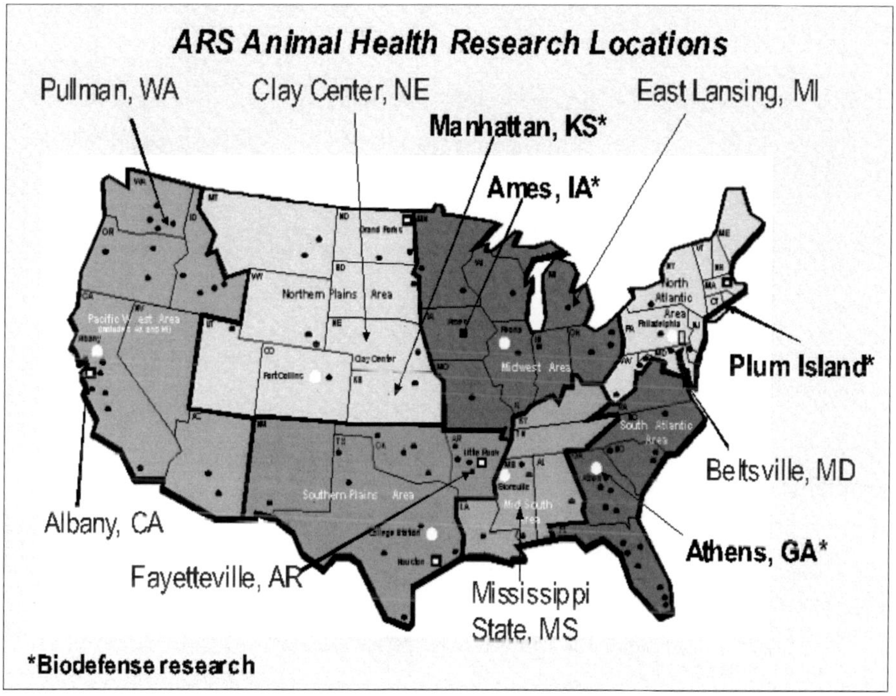

Fig. 1. Location of ARS laboratories conducting animal health research. The locations marked with an asterisk indicate laboratories with biocontainment facilities used to implement the ARS biodefense research program.

Mission and goal

The mission of the National Animal Health Program is to deliver scientific information and tools to detect, control, and when feasible, eradicate animal diseases. The goal is to protect and ensure the safety of the Nation's agriculture and food supply through improved disease detection, prevention, control, and treatment.

Basic and applied research approaches are applied to solve animal health problems of high national priority. Emphasis is given to methods and procedures to control animal diseases through the discovery and development of diagnostics, vaccines, and biotherapeutics.

Strategic objectives

The National Animal Health Program has ten strategic objectives that guide the direction of the program (Table 1). The aim of these objectives is to provide overarching targets to maximize the impact and relevance of the national program. These objectives take into consideration the historical role of the program, research capacity, and expected outcomes.

Table 1: National Animal Health Program strategic objectives

1. Establish ARS laboratories into a fluid and highly effective research network to maximize use of core competencies and resources.
2. Ensure access to specialized high containment facilities to study transboundary, zoonotic and emerging diseases.
3. Integrate the use of animal and microbial genomics research tools.
4. Establish excellence in animal immunology research.
5. Launch a biotherapeutic discovery program providing alternatives to animal drugs.
6. Build technology-driven vaccine and diagnostic discovery research programs.
7. Develop core competencies in field epidemiology and predictive biology.
8. Establish strategic national and international research collaborations.
9. Establish best-in-class training centers for veterinarians and scientists.
10. Develop a model technology transfer program to achieve the full impact of research discoveries.

Action plan and research components

ARS national programs follow a five-year national program cycle that begins with input from stakeholders and the preparation of an action plan (Fig. 2). The action plan provides program direction with the aim of maximizing the use and impact of available resources, and ensures the national program is relevant to the needs of stakeholders. The action plan is organized under research components that include problem statements that define the scope of the action plan.

The National Animal Health Program Action Plan has seven research components. These research components draw upon the relevant expertise within the national program, with the aim of coordinating and integrating that expertise to develop specific useful applications of the knowledge. Projects within the research components take into consideration the need to attract additional federal, university, and industry partners at both the national and international level. The aim of these partnerships is to support component projects. Their inclusion enables and enhances the anticipated products of the component projects. Because a significant number of projects in the animal health research portfolio focus on the discovery of novel technologies, intellectual property strategies are addressed in the project plans developed by ARS scientists to facilitate technology transfers and help foster investments by the private sector in the development of these technologies.

Component 1 of the National Animal Health Action Plan provides the program direction for biodefense research. Outputs under Component 1 are used by Federal and State regulatory agencies for surveillance and to mitigate accidental or potential intentional acts of agroterrorism. The program direction provided to scientists assigned to biodefense research purposely target basic research aimed at increasing our understanding of how disease agents survive outside the host, move between susceptible hosts, infect animals, and how pathogens escape and shed from the host. To improve our response to disease incursions, the program allocates significant resources towards the discovery of veterinary countermeasures for the National Veterinary Stockpile (see below).

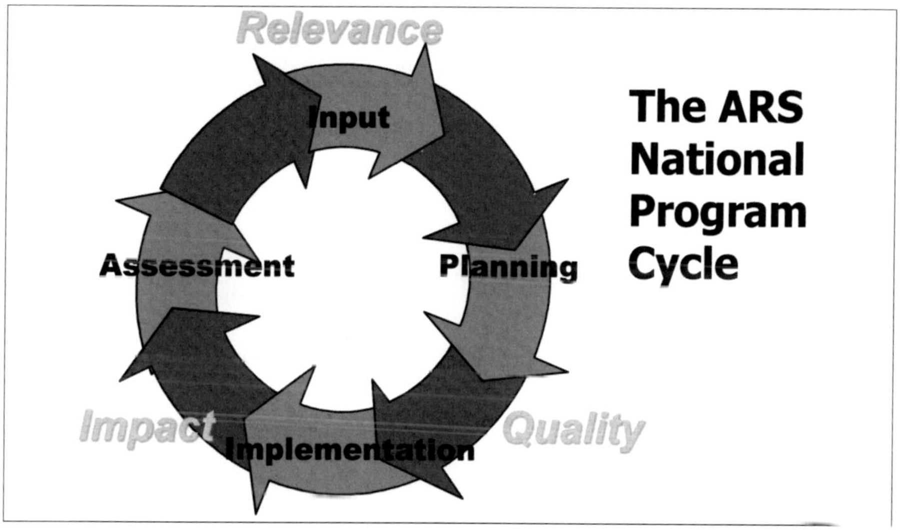

Fig. 2. ARS research is organized around 5-year national program cycles that begin with input from stakeholders. Significant planning is conducted to ensure the national program is relevant to the needs of stakeholders. An action plan is prepared to provide program direction. Prior to the implementation of the research, each project is subjected to a peer-review to ensure quality. The cycle ends with a peer-review of the national program to determine impact.

Biodefense Research

The President's strategy for countering biological threats

The USDA supports the President's Strategy for Countering Biological Threats

[2] through the administration of targeted actions that are linked to the President's seven strategic objectives (Table 2). ARS contributes to these actions by providing scientific information and countermeasures to ensure agricultural production is secure. Many of these countermeasures include diagnostics and vaccines to protect livestock and poultry from biological threats with the potential to cause severe economic consequences, and/or mitigate public health threats in the case of zoonotic diseases. Such biological threats may arise from natural, unintentional, or intentional incursions with domestic or foreign pathogens. Because many dangerous pathogens have the potential to rapidly spread across national (i.e., political) borders, the ARS biodefense research program maintains a global view of the biological threats to food and agriculture with a focus on transboundary animal diseases and emerging diseases

Table 2: President's strategy for countering biological threats

Objective 1: Promote global health security
Objective 2: Reinforce norms of safe and responsible conduct
Objective 3: Obtain timely and accurate insight on current and emerging risks
Objective 4: Take reasonable steps to reduce the potential for exploitation
Objective 5: Expand our capability to prevent, attribute, and apprehend
Objective 6: Communicate effectively with all stakeholders
Objective 7: Transform the international dialogue on biological threats

Transboundary Animal diseases

The health of animals is continuously threatened by diseases naturally or deliberately introduced into a naïve healthy population of productive animals. These diseases vary in the degree of economic loss they cause, their potential to spread, ease of control and ability to eradicate. Many of these diseases are caused by high consequence animal pathogens that are not hindered by international borders and are thus labeled transboundary diseases by the Food and Agriculture Organization (FAO) of the United Nations [3]. Since most of these diseases do not exist in the United States they are commonly referred to as foreign animal diseases. Table 3 identifies the foreign animal diseases for which ARS has active research programs.

Table 3: ARS foreign animal disease research program

African Swine Fever (ASF)
Classical Swine Fever (CSF)
Exotic Blue Tongue Virus (BTV)
Foot-and-Mouth Disease (FMD)
Highly Pathogenic Avian Influenza
Rift Valley Fever (RVF)
Virulent Newcastle Disease

Emerging Diseases

Of particular concern are emerging animal diseases that challenge our disease surveillance systems and our ability to prepare and respond to disease outbreaks. Several new emerging animal disease issues appear every year worldwide. Many factors — globalization of trade, movement of masses of people and agricultural products, changing weather patterns, rapid population growth in cities, intensive agriculture, limited genetic diversity in farm animals, changes in farm practices — are creating new opportunities for the emergence and re-emergence and spread of infectious diseases. Exotic (non-native) organisms, once introduced into the United States, can escalate into an epidemic because of the absence of vaccines or effective drugs, lack of resistance in host animals, and limited resources to effectively manage the spread of such pathogens. Because a large majority of emerging diseases are either vector-borne or zoonotic diseases, ARS takes a One Health approach to planning and implementing research projects through active collaborations with the biomedical research community. ARS collaborative research programs integrate core competencies in infectious diseases, diseases complexes, microbial and animal genomics, pathology, disease detection, and epidemiology to identify new pathogens and predictors of emerging diseases of livestock. Table 4 identifies emerging diseases for which ARS has active research programs.

Table 4: ARS emerging animal disease research program

African Swine Fever***
Avian Influenza*
Blue Tongue Virus***
Brucellosis*
Chronic Wasting Disease (CWD)
Exotic Newcastle Disease*
Foot-and-Mouth Disease (FMD)***
HOBI Virus
Pandemic H1N1**
Rift Valley Fever (RVF)*
Schmallenberg Virus
Tuberculosis*
Variant Swine Influenza Influenza A Virus (vH3N2)*
Highly Pathogenic Porcine Reproductive and Respiratory Disease Syndrome Virus
West Nile Virus*
**Zoonoses **Reverse Zoonosis ***Re-Emerging*

Select Agents

ARS biodefense research activities under Component 1 include research conducted on select agents identified under the Agricultural Bioterrorism Protection Act of 2002 [4]. Select agents pose a severe threat to animal health or animal products. Research on select agents is regulated by the Animal and Plant Health Inspection Service (APHIS) [5] and the Center for Disease Control and Prevention (CDC) [6] and requires high containment laboratories and animal facilities. All of the pathogens included in the ARS foreign animal disease research program (Table 3) are currently on the APHIS agricultural select agent list [7]. The Public Health Security and Bioterrorism Preparedness and Response Act of 2002 set forth provisions for applying for exemptions in the case of investigational products [4]. Obtaining exemption from the select agent program for vaccine strains is a critical milestone in the implementation of a vaccine development plan for foreign animal diseases to remove restrictions that may impede their development in the United States.

Setting Priorities

National Veterinary Stockpile

The National Veterinary Stockpile (NVS) is the nation's repository of vaccines, personal protective equipment, and other critical veterinary products. Homeland Security Presidential Directive 9 (HSPD-9) [8] established the NVS in 2004 to respond to terrorist acts, major disasters, and other emergencies. The directive required the NVS to deploy within 24 hours “sufficient amounts of animal vaccine, antiviral, or therapeutic products to appropriately respond to the most damaging animal diseases affecting human health and the economy…” The directive reflects the national concern that terrorists could simultaneously, in multiple locations, release animal diseases of catastrophic proportions; thus, the need for an NVS that could deliver rapidly large quantities of critical veterinary supplies, equipment, and services to the right place at the right time for as long as necessary. The NVS exists to augment state and local resources in the fight against the APHIS list of the most dangerous animal diseases that could potentially devastate American agriculture.

Developing countermeasures for the NVS presents significant challenges since there is practically no market in the United States for diagnostics and vaccines for pathogens that do not exist in this country. Furthermore, the required product profile for diagnostics and vaccines for foreign animal diseases is more complex than what is required for endemic diseases. For instance, diagnostic tests must be designed to be fit for purpose at every phase of a disease outbreak. Vaccines must be engineered to control and eradicate a disease outbreak, not just prevent clinical disease. Thus, the ideal product profile of vaccines for the NVS requires first and foremost that they prevent transmission to stop the spread of the disease. Since few countermeasures with the ideal profile exist, dedicated research and development programs for the

NVS are paramount. ARS has allocated significant resources under Component 1 of the National Animal Health Program Action Plan to support the research and development needs of the NVS. Success in developing or acquiring relevant diagnostics and vaccines for the NVS requires an in depth assessment of available countermeasures that are either commercially available, or in the pipeline and reachable. To guide research priorities and the acquisition of countermeasures for the stockpile, the NVS has charged ARS with conducting gap analyses for priority setting.

Gap analyses

The specific research priorities and anticipated products from the ARS biodefense research program are derived from gaps analyses conducted with the support and contributions of internationally-recognized disease experts. Many of these gaps analyses were organized by ARS at the request of the National Veterinary Stockpile Steering Committee. The results of these analyses are confidential due to recommendations made in the reports for stockpiling countermeasures. However, some of the reports have been amended for public distribution and can be made available upon request to the Director of the National Veterinary Stockpile, National Center for Animal Health Emergency Management, APHIS.

Interagency Working Group on Foreign Animal Disease Threats

The Interagency Working Group on Foreign Animal Disease Threats (FADT), Committee on Homeland and National Security, National Science and Technology Council (NSTC), developed the 2008-2012 Research and Development Plan, Protecting against High-Consequence Foreign Animal Diseases [9], providing strategic research objectives for implementation and coordination across the federal government. The plan describes progress made to date and identifies new interagency research and development priorities for the period 2008-2012. The research and development priorities follow a One Health approach by taking advantage of interagency integration and collaborations of research in animal, human, and environmental health to enhance the Nation's ability to respond quickly and effectively in the event of a foreign, emerging, or zoonotic animal disease outbreak, whether naturally occurring or intentionally caused. The FADT Interagency Working Group includes four sub-working groups aligned to four major areas:

- The Basic Research Working Group explores opportunities to enhance basic research to better understand disease spread and infection;
- The Modeling Working Group coordinates interagency efforts to enhance the utility of modeling, simulation, and analysis tools for strategic planning and exploration of control strategies and response options for FADT outbreaks at multiple scales;

- The Veterinary Countermeasures Working Group manages efforts to develop diagnostics and vaccines to protect against foreign animal diseases;
- The Depopulation, Decontamination, and Disposal (3D) Working Group coordinates efforts to develop disease-eradication methods, tools, and strategies to facilitate rapid response and recovery.

These sub-working groups enable a coordinated approach to leveraging existing Federal interagency resources to improve the Nation's posture against foreign animal diseases. While each working group has a unique focus, the topics are inextricably intertwined and progress in one area influences the research and development needs of another. Communication and collaboration among the sub-working groups have contributed to the continued success of interagency research and development efforts to address priority research needs to mitigate the threat of foreign animal diseases.

Research Needs

In order to control foreign animal diseases, a wide variety of agent detection platforms need to be developed and validated. Information for designing these platforms requires knowledge of pathogen genomics and proteomics and understanding the evolution and genetic variability of disease agents. There is a dearth of knowledge for many priority foreign animal diseases, such as understanding the host range of pathogens; their primary site of replication; tissue tropism; carrier state; duration and routes of shedding; transmission mechanisms (e.g., vectors, fomites, aerosols); ecology and epidemiology (e.g., wildlife reservoirs). If a disease outbreak should occur in the United States, effective veterinary countermeasures are needed to support suitable control strategies compatible with a short time to recovery to limit the economic impact. There is a need for developing diagnostics, vaccines and biotherapeutics suitable for the NVS. Importantly, integrated approaches for responding to foreign animal disease outbreaks are needed to enhance our capability to regain country disease-free status and retain economic sustainability.

Since no one can predict with certainty the cause of the next pandemic, there is a need to continuously isolate, identify, and characterize pathogen(s) associated with new disease complexes of unknown etiologies. The capabilities to rapidly detect, identify, and characterize new and emerging animal pathogens are paramount and a primary goal of the ARS biodefense research program. Scientists need to conduct animal studies in the relevant host to fulfill Koch's postulate and determine the pathogenesis of monovalent and multivalent infections. Once a new agent is isolated there is a need to sequence partial or complete microbial genomes to identify unique sequences for diagnostic discovery and molecular epidemiology research. Research is needed to understand mechanisms of disease, disease transmission, and host range specificity to determine the prevalence and emerging potential of new diseases. Ultimately research is needed to identify predictors of disease emergence and disease outbreaks. Establishing strategic international research collaborations is critical to address many of these research needs. The following select accomplishments provide examples of research outcomes from the 2006-2011 National Animal Health Program cycle.

Select Accomplishments

Avian Influenza

Avian Influenza (AI) virus is recognized to cause two types of diseases in poultry - a mild form (low pathogenic) that causes respiratory and reproductive problems, and a deadly form (highly pathogenic) that causes systemic disease and high death losses. All AI virus strains that are considered highly pathogenic are reportable to the World Organisation for Animal Health (OIE) and result in immediate closure of international trade markets in poultry and poultry products. International regulations also require reporting of low pathogenic H5 and H7 AI viruses to OIE because of their potential to mutate to the highly pathogenic form of the virus.

The risk of low pathogenic avian influenza viral strains becoming virulent is of significant concern. The use of reverse genetics to engineer avian influenza viruses with a specific sequence has proven to be a valuable tool in examining how these viruses cause disease in poultry. From 1994 to 2006, low-pathogenic H7N2 avian influenza viruses circulated in the live bird markets of northeastern United States. There has been considerable concern that these viruses might change or mutate to the virulent form of the virus. Using reverse genetics techniques, ARS scientists at the Southeast Poultry Research Laboratory in Athens, Georgia, selected a representative H7N2 virus and genetically changed the virus to try and understand the minimum number of nucleotide changes needed for the virus to become virulent [10]. The results of this study showed that the virus needed insertions of amino acids at a key site in the virus (the cleavage site) to become virulent, and simple mutations at the cleavage site by themselves would not make the virus virulent. This study improved our understanding of how AI viruses become virulent and contribute to our ability to predict risks of low-pathogenic viruses changing to the highly pathogenic form in the future.

Classical Swine Fever

Classical Swine Fever Virus (CSFV) is a highly infectious disease of pigs that results in high rates of mortality and morbidity and is a significant foreign animal disease threat to the United States pork industry. Outbreaks in Western Europe have shown that there are critical gaps in the countermeasures available to control this disease. ARS scientists at the Plum Island Animal Disease Center, Greenpoint, New York, have shown that CSFV contains in its envelope three heavily glycosylated (sugar-coated) proteins (E0, E1 and E2), which play a critical role in virulence and immune responses [11]. The role of the glycosylation of these proteins in virulence was established using reverse genetics techniques to alter, delete, and analyze CSFV glycosylation sites. The major outcome from this project is the discovery, mapping, and characterization of viral genetic determinants of virulence and their manipulation, which provide critical information to enable the development of next generation live attenuated vaccine strains.

Exotic Newcastle Disease

Newcastle disease (ND) is an infection of birds caused by a virulent Newcastle disease virus (vNDV) and is a worldwide problem for poultry production. The virulent form of the disease, called exotic ND (END) in the United States, has been eradicated from U.S. poultry. Recent analysis of genomic sequences revealed the existence of two distinct clades: Class I and II. Class I isolates have been recovered primarily from waterfowl and are generally attenuated in poultry. The Class II NDV comprise the vast majority of vNDV and include isolates recovered from poultry, pet, and wild bird viruses. Current weaknesses in the USDA standard real time reverse transcriptase polymerase chain reaction (rRT-PCR) assay to detect class I viruses suggests that viral transmission could occur unnoticed among wild birds and poultry [12]. Although it is impossible to predict which genotypes represent the most significant threat to the U.S. poultry industry, the geographic proximity of vNDV and the large volume of commerce and travel with Mexico and countries of Southeast Asia suggest that further evaluation of the efficacy of current U.S. vaccines and diagnostic assays on emerging viruses are needed. ARS scientists in Athens, Georgia, have developed an alternative rRT-PCR test that can detect a majority of class I isolates from the United States [13]. This test will enable the surveillance of Class I viruses and help identify the genotypes of viruses with the potential to acquire virulence upon replication in poultry, an important step in improving the capacity to prevent future outbreaks.

Foot-and-Mouth Disease

Foot-and-mouth disease (FMD) is one of the most economically and socially devastating diseases affecting animal agriculture throughout the world. Although mortality is usually low in adult animals, millions of animals have been killed in efforts to rapidly control and eradicate FMD. It is well established that the respiratory tract is the most important route of infection of FMD virus (FMDV) in cattle. However, conflicting data from different research groups have implicated regions of either upper respiratory tract (nasopharynx) or lower respiratory tract (lungs) as the primary sites of infection. ARS scientists in Greenpoint, New York, have demonstrated that after aerosol exposure to the virus, the early pathogenesis events involve: 1) primary replication in epithelial cells of the pharyngeal mucosa-associated lymphoid tissue crypts and 2) subsequent widespread replication in pneumocytes in the lungs, which coincides with 3) the establishment of sustained viremia [14,15]. This infection model demonstrated that massive viral amplification occurs in the lungs (with associated shedding to the environment) prior to appearance of the first vesicle. Viremia is established coincidently with further viral amplification in the lungs and at lesion (vesicle) predilection sites. This scientific information is important to the development of FMD countermeasures as it is critical that prophylactic products target these previremic events in the pharynx and lungs. Based on this information, it is speculated that enhancement of mucosal immunity has a high probability of improving protection. Once viremia is established, on an individual animal basis, the battle has already been lost. Continued efforts to improve the understanding of virus host interactions during early phases of infection will greatly contribute to the development of effective tools to block viral infection.

Rift Valley Fever

Rift Valley fever virus (RVFV), a zoonotic insect transmitted virus endemic to Africa and the Arabian Peninsula, causes abortions and high mortality in newborn ruminants. RVFV is continuously re-emerging and could have significant economic and public health outcomes should it be introduced in North America. To determine which arthropods should be targeted for control should RVFV be detected in North America, ARS scientists at the Arthropod-Borne Animal Diseases Research Unit in Manhattan, Kansas, in collaboration with U.S. Army scientists at Fort Detrick, Maryland, evaluated mosquito species from the western, midwestern, and southern United States for their ability to transmit the virus [16]. *Culex tarsalis* mosquitoes transmitted Rift Valley fever virus most efficiently, whereas none of the other *Culex* species tested transmitted the virus as well. Although *Aedes vexans* mosquitoes from Florida and Louisiana were relatively efficient vectors of RVFV, specimens of this species captured in Colorado or California were virtually incompetent, illustrating the need to evaluate local populations for the ability to transmit a pathogen. This was the first comprehensive vector competence study of U.S. mosquitoes for RVFV and provides key targets for vector control should an outbreak of Rift Valley fever virus ever occur in the United States.

Swine Influenza Virus

Soon after the emergence of the pandemic H1N1 Influenza A virus in April 2009, ARS scientists at the National Animal Disease Center, Ames, Iowa, began research using virus samples provided by the CDC. The first step was to address the safety of pork. Thirty-five-week-old cross-bred pigs from a herd free of swine influenza virus were inoculated to determine the susceptibility of swine to the human virus and also address whether meat, blood, and tissue from pigs infected with the pandemic 2009 A/H1N1 influenza virus are free of the infectious virus [17]. The pigs were observed daily for clinical signs of disease, and nasal swabs and fresh samples from lung, tonsil, inguinal lymph node, liver, spleen, kidney, skeletal muscle (ham), and colon contents were tested by the most sensitive virus detection assays at three, five, and seven days post-infection. Live 2009 A/H1N1 influenza virus was detected only in the respiratory tract of infected pigs, and the virus did not appear to spread and replicate in other tissues. This information was used by the pork industry during the pandemic H1N1 outbreak to communicate to consumers that pork was safe to eat.

Conclusion

ARS is the principal in-house research agency of the USDA. It is one of the four component agencies of the Research, Education, and Economics mission area. ARS conducts research to find solutions to problems of high national priority, including biodefense research to protect the United States from biological threats. The ARS biodefense program specifically targets the research needs of USDA action and regulatory agencies and a number of other Federal agencies within the federal government. The primary mission of the ARS National Animal Health Program is to deliver scientific information and tools to detect, control, and eradicate animal diseases. ARS supports the President's National Strategy for Countering Biological Threats through the administration of targeted actions. ARS biodefense research activities also include research conducted on select agents identified under the Agricultural Bioterrorism Protection Act of 2002. To improve the USDA's response to disease incursions, the ARS biodefense research program allocates significant resources towards the discovery of veterinary countermeasures for the U.S. National Veterinary Stockpile. Research priorities and anticipated products from the ARS biodefense research program are derived from gaps analyses conducted with the support and contributions of internationally-recognized disease experts. ARS is an active participant in the NSTC Interagency Working Group on Foreign Animal Disease Threats, providing strategic research objectives for implementation and coordination across the federal government. This Interagency Working Group provides a coordinated approach to leveraging existing Federal interagency resources to address the Nation's research needs for foreign animal diseases. Establishing strategic international research collaborations is critical to address many of these research needs. Select research accomplishments from the 2006-2011 National Animal Health Program cycle highlight the importance and scope of the ARS biodefense research program.

Ackowledgments

The author thanks ARS scientists for their dedication to the animal health biodefense research program, and U.S. Government stakeholders at APHIS, DHS, CDC, and the Department of State, whose support is critical to the mission and success of the ARS National Animal Health Program.

REFERENCES

1 United States Department of Agriculture, Agricultural Service, Research, National Programs: www.ars.usda.gov/research/programs.htm.

2 White House, National Security Council, National Strategy for Countering Biological Threats, November 2009: www.whitehouse.gov/sites/default/files/National_Strategy_for_Countering_BioThreats.pdf.

3 FAO, Agriculture Department, Animal Production and Health Division, Programs, Transboundary Animal Diseases: http://www.fao.org/ag/againfo/programmes/en/empres/diseases.asp.

4 Public Health Security and Bioterrorism Preparedness and Response Act of 2002, Title II, Enhancing Controls on Dangerous Biological Agents and Toxins, Subtitle B, Agricultural Bioterrorism Protection Act 2002: www.aphis.usda.gov/programs/ag_selectagent/downloads/PL107-188.pdf.

5 United States Department of Agriculture, Animal and Plant Inspection Service, Agricultural Select Agent Program: www.aphis.usda.gov/programs/ag_selectagent/.

6 Centers for Disease Control and Prevention, Office of Public Health Preparedness and Response, Regulations of Select Agents and Toxins: www.cdc.gov/phpr/dsat.htm.

7 United States Department of Agriculture, Animal and Plant Inspection Service, Agricultural Select Agent Program, Select Agents and Toxins List: www.aphis.usda.gov/programs/ag_selectagent/ag_bioterr_toxinlist.shtml.

8 Homeland Security Presidential Directive-9 (HSPD-9), "Defense of United States Agriculture and Food," January 30, 2004, www.fas.org/irp/offdocs/nspd/hspd-9.html.

9 NSTC Subcommittee on Foreign Animal Disease Threats. Committee on Homeland and National Security. Protecting Against High-Consequence Animal Diseases: Research & Development Plan for 2008-2012. January 2007. http://www.whitehouse.gov/sites/default/files/microsites/ostp/fadt_rd_16_feb_2007.pdf.

10 Lee C, Lee Y, Senne DA, Suarez DL: Pathogenic potential of North American H7N2 avian influenza virus: A mutagenesis study using reverse genetics. Virology 2006;353:388-395.

11 Sainz IF, Holinka LG, Lu Z, Risatti GR, Borca MV: Removal of a N-linked glycosylation site of classical swine fever virus strain Brescia Erns glycoprotein affects virulence in swine. Virology 2008;370(1):122-129.

12 Kim LM, Afonso CL, Suarez DL: Effect of probe-site mismatches on detection of virulent Newcastle disease viruses using a fusion-gene real-time reverse transcription polymerase chain reaction test. J Vet Diagn Invest 2006;18:519-528.

13 Kim LM, King DJ, Suarez DL, Wong CW, Afonso CL: Characterization of Class I Newcastle disease virus isolates from Hong Kong bird markets and detection using real-time reverse transcription PCR. J Clin Microbiol 2007;45:1310-1314.

14 Arzt J, Gregg DA, Clavijo A, Rodriguez LL: Optimization of immunohistochemical and fluorescent antibody techniques for localization of Foot-and-mouth disease virus in animal tissues. J Vet Diagn. Invest 2009;21:779-792.

15 Arzt J, Pacheco JM, Rodriguez LL: The early pathogenesis of foot-and-mouth disease in cattle after aerosol inoculation: identification of the nasopharynx as the primary site of infection. Vet Pathol 2010;47(6):1048-1063.

16 Turell M, Bennett KE, Wilson W: Potential for North American mosquitoes to transmit Rift Valley fever virus. J Med Ent 2010;47:884-889.

17 Vincent AL, Lager KM, Harland M, Lorusso A, Zanella E, Ciacci-Zanella JR, et al: Absence of 2009 pandemic H1N1 influenza A virus in fresh pork. PLoS ONE 2009;4(12):e8367.

Cyril G. Gay, DVM, Ph.D, Senior National Program Leader, Animal Production and Protection, Office of National Programs, Agricultural Research Service, United States Department of Agriculture

Tel: 301 504-4786

Email: cyril.gay@ars.usda.gov

Roth JA, Richt JA, Morozov IA (eds): Vaccines and Diagnostics for Transboundary Animal Diseases. Dev Biol (Basel). Basel, Karger, 2013, vol 135, pp 39-52.

Biologics Industry Challenges for Developing Diagnostic Tests for the National Veterinary Stockpile

J.M. Hardham, C.M. Lamichhane

Global Biologicals Research Zoetis Inc., Kalamazoo, MI, USA

Key words: veterinary, stockpile, diagnostic, industry, challenge

Abstract: Veterinary diagnostic products generated ~$3 billion US dollars in global sales in 2010. This industry is poised to undergo tremendous changes in the next decade as technological advances move diagnostic products from the traditional laboratory-based and handheld immunologic assays towards highly technical, point of care devices with increased sensitivity, specificity, and complexity. Despite these opportunities for advancing diagnostic products, the industry continues to face numerous challenges in developing diagnostic products for emerging and foreign animal diseases. Because of the need to deliver a return on the investment, research and development dollars continue to be focused on infectious diseases that have a negative impact on current domestic herd health, production systems, or companion animal health. Overcoming the administrative, legal, fiscal, and technological barriers to provide veterinary diagnostic products for the National Veterinary Stockpile will reduce the threat of natural or intentional spread of foreign diseases and increase the security of the food supply in the US

Introduction

In the United States (US), animal agriculture is valued at $165 billion and is a principal source of food, a major source of livelihood for Americans, and a major contributor to US agricultural exports [1]. The globalization of the livestock industry, which has made large-scale and rapid movement of animals commonplace, increases the opportunities for rapid global transfer of infectious disease agents to naïve animal populations [2-4]. In addition, there is a constant risk of infectious disease agent misuse for political or sociological gain [5]. The National Research Council stated that the US is vulnerable to agricultural bioterrorism [6] and the National Academies of Science stated that "the security of this multibillion-dollar enterprise and of the food system to which animal agriculture is intricately connected remains vulnerable to disease threats, whether intentionally or naturally introduced" [1]. Given the risks to the agricultural sector, there is a need to protect this vital national asset from the threats of foreign animal diseases (FADs), new or emerging infectious disease agents, and bioterrorism.

The US government is charged with protecting the citizens, animals, and plants of this country from the threat of foreign diseases, whether they reach our soil naturally or via intentional release. Homeland Security Presidential Directive 5 (Management of Domestic Incidents) (HSPD-5) [7] identifies the Secretary of Homeland Security as the principal Federal official responsible for domestic incident management. As such, the Department of Homeland Security (DHS) and the United States Department of Agriculture (USDA) have the responsibility to develop and stockpile medical countermeasures (MCMs) to detect, prevent, and treat FADs.

HSPD-9 (Defense of United States Agriculture and Food) [8] established the national policy to defend the US agriculture and food system against terrorist attacks, major disasters, and other emergencies. HSPD-9 called for the establishment of a stockpile of veterinary MCMs to aid in the protection of the nation's food supply. The proposed stockpile would maintain sufficient amounts of MCMs that would be capable of being deployed within 24 hours of a declared veterinary emergency. The USDA Animal and Plant Health Inspection Service (APHIS) Veterinary Services established the National Veterinary Stockpile (NVS) in 2004 [9] and currently provides for its maintenance, replenishment, and updates. The mission of the NVS is to provide the veterinary MCMs (diagnostic tests, animal vaccines, antivirals, or therapeutic products), supplies, equipment, and response support services that States, Tribes, and Territories need to respond to damaging animal disease outbreaks. The USDA APHIS National Center for Animal Health Emergency Management (NCAHEM) develops strategies and policies for effective incident management and helps to coordinate incident responses. As a liaison to outside emergency management groups, NCAHEM ensures that emergency management policies, strategies, and responses are current with national and international standards [10].

The National Science and Technology Council Foreign Animal Disease Threats Working Group (FADT WG) provides oversight and guidance to the NVS regarding disease threats and recommendations for MCM research, development, and procurement. The Working Group has identified key technologies needed to protect the US agricultural system; the supporting research to develop these technologies; and made recommendations for strategic investments [11]. The Research and Development (R&D) plan identifies the top priority disease agents (Table 1) and states that "better, more effective vaccines, biological, and diagnostics will allow us to detect and respond to disease more quickly and to more effectively stop or slow the course of a disease." The plan identified three major areas of research, development, test, and evaluation (RDT&E): (1) Diagnostic technologies to provide tools for detection and surveillance and surge capacity during response and recovery; (2) Vaccines and immunomodulators needed for an effective response to an incursion of the disease; and (3) Basic research on disease epidemiology and disease pathogenesis to provide a greater understanding of disease spread and countermeasure potentials.

Table 1. USDA APHIS list of the 17 most damaging animal disease threats

Priority	**foreign animal disease**
1	Highly Pathogenic AI (a, b)
2	Foot-and-Mouth Disease (b)
3	Rift Valley fever (a)
4	Exotic Newcastle Disease (b)
5	Nipah and Hendra virus (a)
6	Classical swine fever (b)
7	African swine fever (b)
8	Bovine spongiform encephalopathy
9	Rinderpest (b)
10	Japanese encephalitis (a)
11	African horse sickness
12	Venezuelan equine encephalitis (a)
13	Contagious bovine pleuropneumonia
14	*Ehrlichia ruminantium* (Heartwater)
15	Eastern equine encephalitis (a)
16	*Coxiella burnetii* (a)
17	Akabane virus
(a) Zoonotic Disease	
(b) FBI Pathogen of concern	

This FADT WG R&D plan falls short, however, in that it does not provide an effective strategy to transition basic research technology developments from Academic or Government laboratories into industrial development and production. The President's Council of Advisors on Science and Technology (PCAST) recently released the "Report to the President on Capturing Domestic Competitive Advantage in Advanced Manufacturing" [12]. The report states, "Maintaining technological superiority in advanced manufacturing is a national security issue and is critically important for sustaining US global competitiveness. A strong manufacturing sector not only ensures a ready supply of defense and commercial goods and services, but also ensures the integrity of these goods, especially electronics and other mission critical items. However, US national security is not limited to the products and technologies that are required for national defense; it also entails the products and technologies required for our nation's energy security, food security, heath security, cybersecurity and economic security." It further states that "The US, and the world, is witnessing the global importance of food security; advanced manufacturing will enable the world to feed its growing population in a sustainable, efficient manner…" Annex 1 of the report also states that "Food security is also going to be a key concern of the future, where again bio manufacturing will play a critical role."

Despite the importance of developing and producing MCMs for FAD threats, there exists a gap between basic research and advanced product development. The PCAST Report Annex 2 identifies that "US universities receive a great deal of Federal funding for basic and applied research" but "only a small number of discoveries and findings are translated into new products." This is due to the fact that "The business sector performs the largest portion of US R&D work, using internal resources." The decision to invest in research and development of new animal health pharmaceutical products, vaccines, and diagnostic assays by the Animal Health Industry is based on potential of the products to generate sales in a given target market. The decision process involves the assessed potential market size and potential product sales being weighed against the cost of product development. The potential return on investment amongst various potential projects is then compared so that the limited corporate resources may be best applied toward projects that have the highest chances of success in generating future earnings and providing a return for investors. This decision process works very well for diseases that exist globally or in regions of high economic importance. Animal Health companies struggle with the decision to progress projects for emerging infectious disease agents or FAD agents. The difficulty in assessing these types of projects is that there are no or very small markets available for sales of the product and therefore result in a negative return on investment. In other words, it would cost more to develop the product than the product would sell.

Industry Process

The global veterinary diagnostic market represents a fast growing and highly technical market with approximately $3 Billion US dollars of sales in 2010 (Fig. 1). The development of vaccines and diagnostics in animal health progresses in much the same fashion as it does in human health, although the time from concept to product is shorter for animal health products. Fig. 2 shows typical product progression paths for animal health vaccine and diagnostic products. Product concepts are evaluated based on their scientific/technical merits and on their potential to provide a financial return on the investment. Since the companies that discover, develop, and license animal health products are profit driven, they must utilize their scarce resources on projects and programs with the highest potential to yield products that will return a profit to the company. Many factors are considered when evaluating the potential return on investment of a product. These include the market size, the market distribution, the owners willingness to treat, the cost of the product, the safety and efficacy of the product, and ability of the product profile to satisfy a market need. Once a program has been endorsed, it is allocated resources (financial, physical, and human). Both vaccine and diagnostic programs typically move from a discovery phase (where numerous targets, assay types, antigens, and formulations are evaluated and proof of concept is demonstrated) into a development phase (where a lead product candidate is scaled-up, validated, and prepared for regulatory filing). One of the key aspects of an efficient product development program is the establishment of a product profile early in the discovery phase.

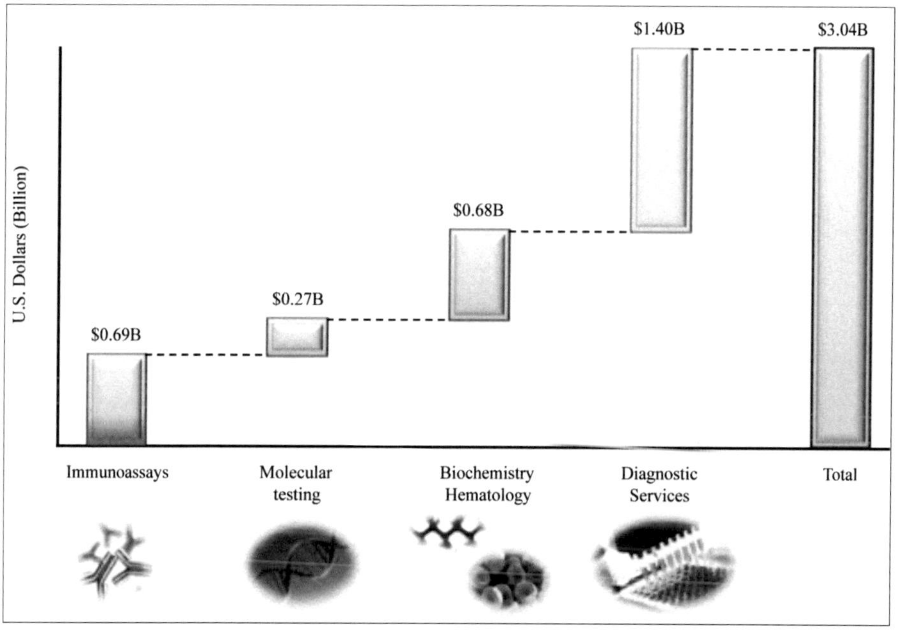

Fig. 1. World market for veterinary diagnostics. Values are shown in billions of US dollars. The global sales figures (grey box) are broken down by technology type. Immunoassays (red) are made up of test kits and reagents such as ELISAs for food animals and point-of-care diagnostic kits for companion animals. Molecular test (blue) are DNA or RNA detection kits and reagents. Biochemistry/hematology (purple) are blood and blood product kits, analyzers, and reagents. Diagnostic services (green) are the fastest growing segment that is composed of laboratory diagnostic services to conduct and analyze diagnostic tests. Data source: Kalorama:"The world market for veterinary diagnostics", 2012; BCG analysis 2010.

There are numerous opportunities for synergy between vaccine and diagnostic product development programs targeting the same disease agent. This is especially true when products are required that allow for the differentiation between infected animals and vaccinated animals (DIVA). For example, development of a recombinant capsid vaccine targeting a virus allows for the development of a compatible diagnostic assay targeting a non-structural protein (Fig. 3). In this case, animal sera from the various safety and efficacy studies of the vaccine program may be utilized for assay development and refinement in the diagnostic assay development program. Hence, co-development of vaccine and compatible diagnostic products saves both time and money.

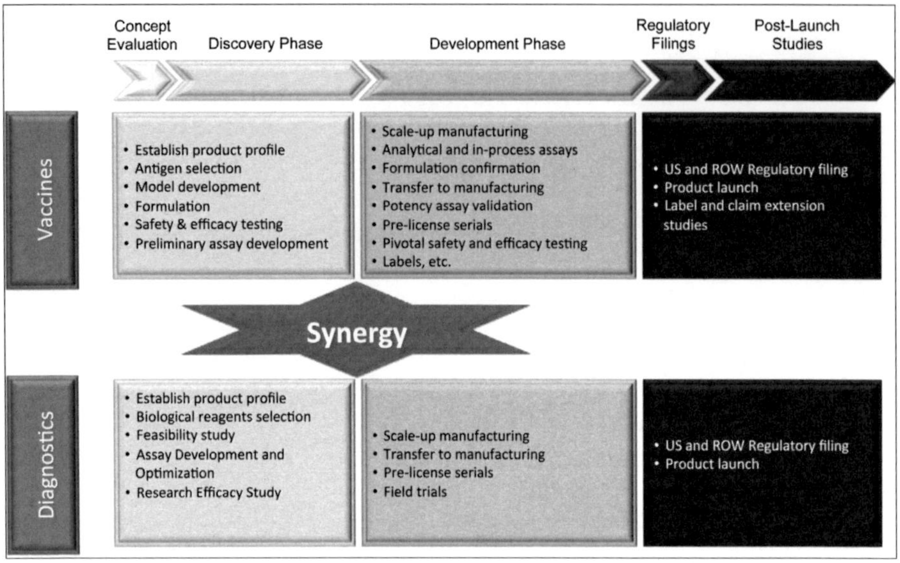

Fig. 2. Typical vaccine and diagnostic product progression paths for the animal health industry.

Expanding Role of Diagnostic Assays

FADs pose a great threat to the poultry, equine, cattle, and swine industries worldwide. The demands for development of rapid, accurate, and sensitive diagnostic methods for detecting and differentiating pathogens are increasing. These methods are the basis for treating, controlling, and eradicating infectious diseases of veterinary importance.

Classical pathogen detection and identification are based on culture methodology, biochemical tests, microscopy, and conventional immunoassays. Culture methodologies and conventional immunoassays are still a core technology in many clinical laboratories because they provide important information about the viability of the pathogen, immune status of animals, and susceptibility to therapeutic agents. Recovery of the live agent and measuring immune status of animals remains a very slow, technically demanding, and expensive diagnostic tool.

The development of rapid, accurate, and sensitive diagnostic methods for detecting disease pathogens are fundamental for treating, controlling, and eradicating infectious disease. Rapid scientific and technological developments have revolutionized the possibilities of diagnosing and treating disease at the clinical and farm levels.

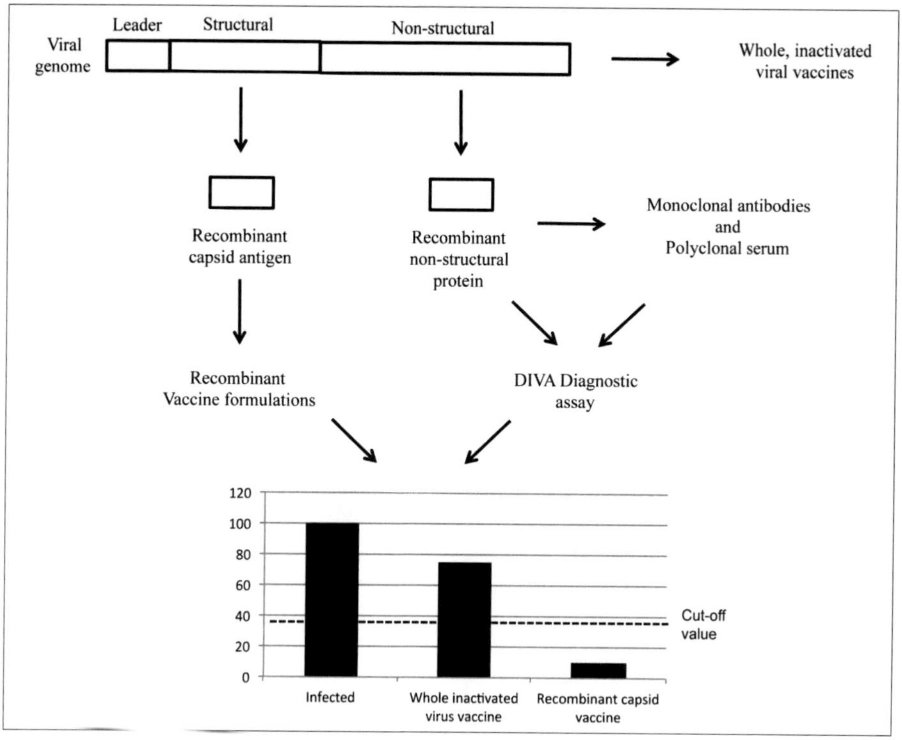

Fig. 3. Depiction of recombinant vaccine and compatible DIVA diagnostic product approaches.

Researchers now have tools for observation at the atomic level, to sequence entire genomes, and to understand the molecular basis of disease. In addition, recent biotechnological developments in microtechnologies and nanotechnologies have led to the proliferation of new, rapid diagnostic tests that hold the promise for improved health management and rapid control of infectious diseases of animals. Emerging technologies are rapidly being adapted to improve our ability to detect new and re-emerging etiological agents, to provide rapid, accurate diagnostic tests for previously difficult infections, and to develop new response-based paradigms for diagnostic testing. Molecular tools in particular are advancing toward laboratory-on-chip and fully integrated technologies that are capable of same-day detection and pathotyping of agents in clinical specimens. New immunodiagnostics technologies that are rapid, portable, low cost, and of greater sensitivity/specificity are aiding veterinarians in early disease detection in a point-of-care setting. Highly sensitive and specific diagnostic tests are no longer confined to fixed laboratories. Rather, they are evolving to be field-based assays that are capable of wirelessly streaming data to clinicians in real time.

Challenges and Recommandations

The development and ultimate licensure of vaccines and compatible DIVA diagnostic assays for animal health use is complex and costly. It is even more challenging when the products are targeting foreign animal disease threat agents. The challenges that are faced by the animal health industry in developing vaccines and diagnostic products for foreign animal disease threats can be broken down into four categories (process, legal, fiscal, and technological). These challenges are depicted in Fig. 4.

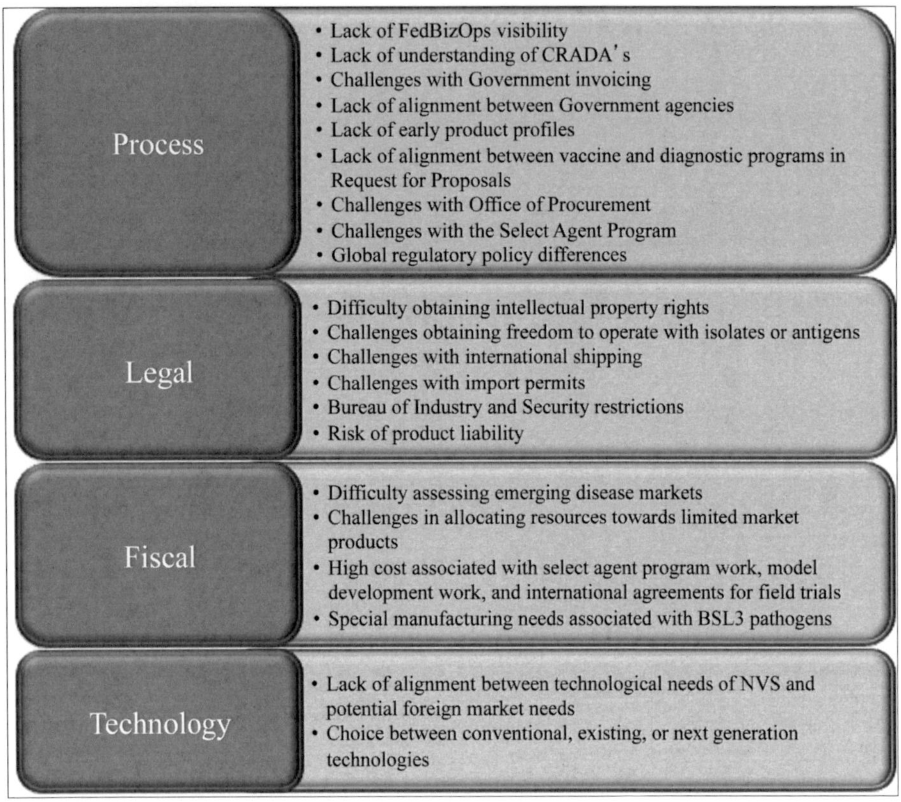

Fig. 4. Examples of animal health industry challenges in foreign animal disease product development.

Process

The mechanisms by which the animal health industry interacts with the US Government during the development of vaccines and diagnostics for FADs may inherently create challenges and introduce additional corporate risk into the process. Often times, the key scientific decision makers within large animal health companies are unaware of the opportunities that the US Government has to offer due to a lack of visibility of Requests for Information (RFI) or Requests for Proposal (RFP) published on FedBizOps. Many companies have a business group that is charged with scanning federal publications. However, when an opportunity is identified, there is often very little window of opportunity to respond since the opportunities may only have a 30- or 60-day solicitation window. A targeted distribution of RFIs and RFPs via The Animal Health Institute [13] or The International Federation for Animal Health [14] may improve visibility and allow for a longer response time.

The usage of a product often will dictate how that product is developed. There are four distinct phases of a response to a FAD incident – surveillance, response, recovery, and post-response surveillance. The product profiles for a product targeting a specific FAD agent may be different for each of these phases. As such, the development plan and costs may vary. Ambiguities in the product profile at the start of the product development process lead to increased risk for the program. This risk may be reduced by early development and communication of the detailed product profile.

The fiscal challenges of developing products targeting FAD agents may often be reduced if the product may be sold to overseas markets where the FAD is endemic. The challenge for the US product is that the regulatory requirements between countries are not synchronized. Hence, a product developed for an overseas market may or may not be licensable in the US. The increased efforts to globally harmonize regulatory requirements will decrease this challenge.

Legal

The industrial decision to develop and market diagnostic assays for veterinary infectious disease agents is a profit-based process. The profitability of a diagnostic product is based on the net sales less the cost of developing, producing, and delivering the product to the market. If the reagents required to produce a diagnostic product carry too many royalties (due to in-licensing of intellectual property (IP)), then the product may cost too much to produce and therefore, significantly reduce or even eliminate its profitability. This is especially so in the development of diagnostic assays for FADs where the market size is extremely small. It is key to the animal health industry to have access to FAD antigens and/or isolates that are either free of IP or have minimal royalties associated with them.

In addition to IP, there are a number of other legal aspects of the development of FAD diagnostic assays that impact the ability to progress these programs. International shipping and importation of FAD agents and/or components creates barriers to rapid development timelines. For global animal health companies, the US Bureau of Industry and Security (BIS) restricts the dissemination of information regarding certain aspects of manufacturing products relating to select agents [15]. These additional challenges increase cost and risk in the development process.

Fiscal

Animal Health companies are challenged in evaluating potential products targeting FAD threat agents since the potential domestic market is either non-existent or is extremely limited [16]. A key piece of information that Animal Health companies utilize in the decision to initiate a product development program is the Net Present Value (NPV) of a potential product. Basically, the NPV is potential profit of a product less the investment that is required to develop it, produce it, and distribute it. Without a significant domestic market to sell products to, the NPV of FAD MCMs is frequently very small or is negative. It is difficult to justify the utilization of scarce resources towards potential products with negative NPVs when other potential projects may yield significantly higher return on investment. Thus, there is a gap between academic research on FADs and product licensure and manufacture (Fig. 5). This gap is often referred to as the "valley of death" for MCMs. There are multiple government reports that reflect this challenge. The June 2011 PCAST Advanced Manufacturing Report [17] stated that "Individual companies cannot justify the investment required to fully develop many important new technologies or to create the full infrastructure to support advanced manufacturing. Private investment must be complemented by public investment. Key opportunities to overcome market failures include investing in the advancement of new technologies with transformative potential, supporting shared infrastructure, and accelerating the manufacturing process through targeted support for new methods and approaches."

In a February 2011 Report to Congress, the Congressional Research Service stated that, "the government needs to take a larger role in developing countermeasures in areas where the private sector has failed to produce desired countermeasures" [18]. Hence, the DHS Science and Technology (DHS S&T) and the USDA should assist the animal health industry in the development of FAD MCMs by providing codevelopment funding similar to that done by The Department of Defense (DoD) and The Department of Health and Human Services (HHS) in the human MCM development programs for chemical, biological, radiological, and nuclear (CBRN) agents. The 2004 Office of Science and Technology Policy's blue ribbon panel on the threat of biological terrorism directed against livestock [19] recommended the creation of "an animal equivalent of Bio-Shield within DHS or elsewhere in the federal government" in order to provide funding for codevelopment of MCMs targeting FAD threat agents. Codevelopment funding would offset the industrial product development investment and positively impact the NPV of FAD MCMs, thereby increasing the ability of animal health companies to justify FAD MCM development programs.

The 2012 National Strategic Plan for Advanced Manufacturing [20] also calls for the Federal Government to play an important role in investing in R&D to help foster the development of advanced manufacturing processes and products. DHS S&T and the USDA should investigate the potential of partnering with the DoD and/or HHS on the Medical Countermeasure Initiative [21, 22] in order to provide

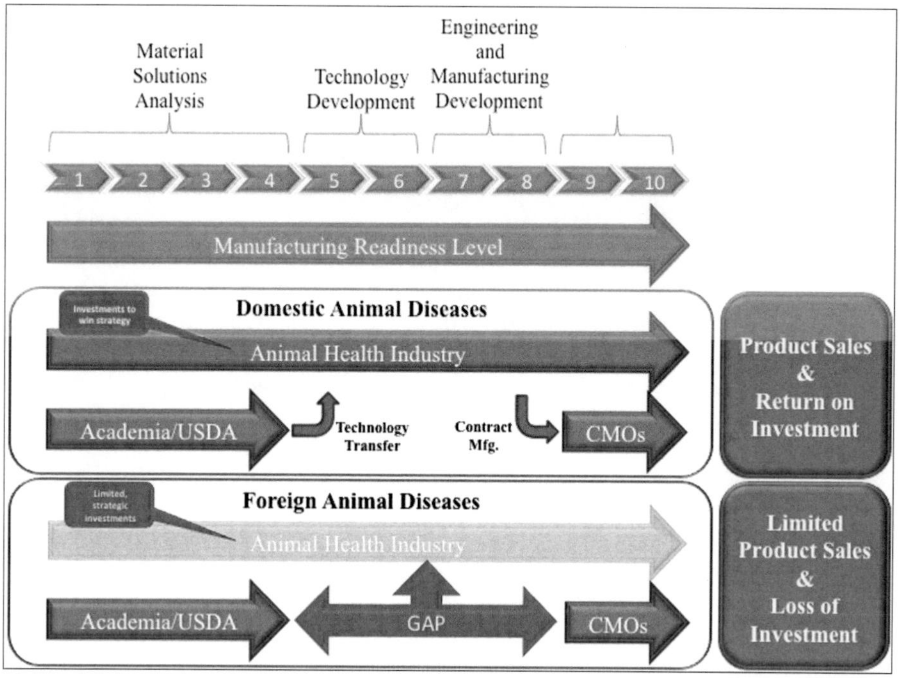

Fig. 5. Graphical representation of the "Valley of Death" for product development of foreign animal disease medical countermeasures.

Advanced Development and Manufacturing capabilities to their performers in order to shepherd vaccine and diagnostic projects across the "valley of death" for FAD MCMs. In addition, the DoD and HHS MCM development programs have been confined into the Integrated Portfolio for CBRN Medical Coutermeasures. DHS S&T should also consider either coordinating or combining portfolios as many of the agents are of shared concern.

Technological

The rapid evolution of diagnostic assay technology creates both opportunities and challenges in the development of diagnostic assays for FAD threat agents. The choice of diagnostic platforms to utilize for FAD assay development is complex. The use of existing diagnostic platforms reduces product development cost and risk, yet may result in a product that is out-dated in terms of future veterinary practices. The use of next-generation diagnostic platforms such as distributed point-of-care diagnostic, portable molecular, and nanotechnology platforms may increase the accuracy, sensitivity, and flexibility of FAD diagnostic assays, yet they increase the cost and risk of product development programs. Product development for foreign markets may influence the choice of diagnostic platforms. However, the product profile for foreign market products and domestic products for the NVS are frequently quite different. The potential of a single product to satisfy the product profile needs of both of these markets is quite small. With the FAD MCM development programs having small or negative NPVs, it is challenging to argue for investment in multiple products and next-generation technologies.

Conclusions

The threat of FADs is a National Security issue for the US. The development, licensure, and stockpiling of FAD MCMs is of critical national importance. The administrative, legal, fiscal, and technological hurdles in FAD MCM product development have created the "valley of death" for new products that is quite difficult to overcome. New and innovative federal solutions and partnerships with industry may help to reduce these challenges and minimize the gap that exists between basic research and product licensure. With the fiscal realities and budget reductions in the government arena, interagency partnerships should be pursued in order to increase efficiency and reduce duplication in infectious disease MCM development.

The development and licensure of new diagnostic assays is only one part of the diagnostic solutions for FAD threats (Fig. 6). The ability to manage the information generated by diagnostic assay results using regional or global biosurveillance systems that incorporate human, veterinary, and environmental data plays a key part in the ability to rapidly address impending FAD threats. The ability to detect the emergence or spread of FADs in the regions where such diseases are endemic or into neighboring regions as early as possible should be a national priority [23]. As the recent case of Schmallenberg virus demonstrates, early detection and characterization of an emerging disease allows for rapid development of diagnostic tools and preventative biologicals [24]. DHS S&T should investigate partnering with other government agencies such as HHS and DoD in ongoing biosurveillance programs to bolster the Nation's early warning systems for veterinary disease outbreaks.

Solving the "valley of death" challenges in development of MCMs for FADs will take the concerted effort of Congress, multiple Federal agencies, and the Animal Health Industry. The continued safety of the US agricultural sector is dependent on a robust suite of MCMs for FADs that provide capabilities to rapidly and accurately detect the presence of a wide variety of FADs, ascertain the spread of the FADs, and respond with efficacious preventative and therapeutic products.

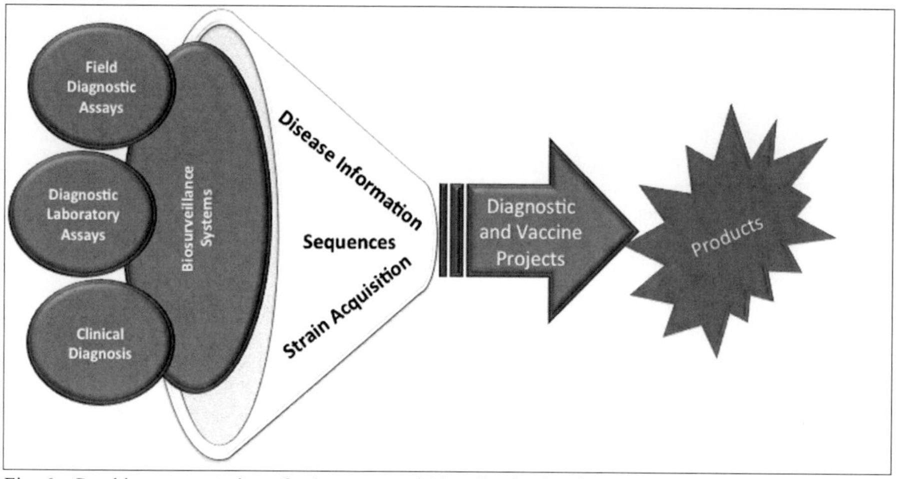

Fig. 6. Graphic representation of substrate acquisition for the development of medical countermeasures (diagnostics and vaccines) targeting foreign animal diseases.

References

1 McElwain T, Connell N, Hennessy D, King LJ, Le Duc J, Maclachlan NJ, et al: Meeting critical laboratory needs for animal agriculture: Examination of three options. National Academies Press. 2012.

2 Brown C: Emerging diseases: The global express. Vet Pathol 2010;47:9-14.

3 Cutler SJ, Fooks AR, van der Poel WHM: Public health threat of new, reemerging, and neglected zoonoses in the industrialized world. Emerg Infect Dis 2010;16:1-7.

4 Tomley FM, Shirley MW: Livestock infectious diseases and zoonoses. Philos Trans R Soc London B Biol Sci 2009;364:2637-2642.

5 Gerstein DM: Bioterror in the 21st century: Emerging threats in a new global environment. Naval Institute Press. Annapolis, MD. 2009.

6 NRC Report. Countering Agricultural Bioterrorism. National Academies Press, 2002.

7 Homeland Security Presidential Directive / HSPD-5. Management of Domestic Incidents. February 28, 2003. http://www.gpo.gov/fdsys/pkg/PPP-2003-book1/pdf/PPP-2003-book1-doc-pg229.pdf.

8 Homeland Security Presidential Directive / HSPD-9. Defense of United States Agriculture and Food. anuary 30, 2004. http://www.gpo.gov/fdsys/pkg/PPP-2004-book1/pdf/PPP-2004-book1-doc-pg173.pdf

9 United States Department of Agriculture, Animal and Plant Health Inspection Service, National Veterinary Stockpile. http://www.aphis.usda.gov/animal health/emergency_management/nvs.shtml

10 United States Animal Health Report. Chapter 2. Prevention and Preparedness. http://www.aphis.usda.gov/animal_health/animal_health_report/

11 Subcommittee on foreign animal disease threats. Protecting Against high consequence animal diseases: Research and development plan for 2008-2012. http://www.whitehouse.gov/sites/default/files/microsites/ostp/fadt_rd_16_feb_2007.pdf

12 President's Council of Advisors on Science and Technology. Report to the President on capturing domestic competitive advantage in advanced manufacturing. July 2012. http://www.whitehouse.gov/sites/default/files/microsites/ostp/pcast_amp_steering_committee_report_final_july_27_2012.pdf

13 Animal Health Institute. http://www.ahi.org

14 The International Federation for Animal Health. http://ifahsec.org

15 The U.S. Bureau of Industry and Security. http://www.bis.doc.gov/

16 Gay CG, Salt J, Balaski C: Challenges and opportunities in developing and marketing vaccines for OIE list A and emerging animal diseases, in Brown F and Roth J (ed): Vaccines for OIE List A and Emerging Animal Diseases. Dev Biol. Basel, Karger, 2003, vol 114, pp 243-250.

17 President's Council of Advisors on Science and Technology. Report to the President on ensuring American leadership in advanced manufacturing. June 2011. http://www.whitehouse.gov/sites/default/files/microsites/ostp/pcast-advanced-manufacturing-june2011.pdf

18 Gottron F, Shea DA: Federal efforts to address the threat of bioterrorism: Selected issues and options for Congress. Congressional Research Service, 7-5700, R41123. www.crs.gov

19 Kelly TK, Chalk P, Bonomo J, Parachini J, Jackson BA, Gary C: The Office of Science and Technology Policy blue ribbon panel on the threat of biological terrorism directed against livestock, Conference proceedings. RAND Corporation, 2004, pp1-24.

20 National Science and Technology Council Committee on Technology. A National Strategic Plan for Advanced Manufacturing. February 2012.
http://www.whitehouse.gov/sites/default/files/microsites/ostp/iam_advancedmanufacturing_strategicplan_2012.pdf

21 The Department of Defense Medical Countermeasure Initiative. http://www.acq.osd.mil/cp/

22 The Department of Health and Human Services Medical Countermeasure Initiative. http://www.phe.gov/Preparedness/mcm/mcmreview-1yr/Pages/innovation.aspx

23 Merianos A: Surveillance and response to disease emergence. Curr Top Microbiol Immunol 2007;315:477-509.

24 Hoffmann B, Scheuch M, Hoper D, Jungblut R, Holsteg M, Schirrmeier H, et al: Novel Orthobunyavirus in cattle, Europe, 2011. Emerg Inf Dis 2012;18:469-472.

Abbreviations

APHIS:	Animal and Plant Health Inspection Service
BIS:	US Bureau of Industry and Security
CBRN:	Chemical, Biological, Radiological, and Nuclear
DHS S&T:	Department of Homeland Security Science and Technology
DHS:	Department of Homeland Security
DIVA:	Differentiation between Infected animals and Vaccinated Animals
DoD:	Department of Defense
FADs:	Foreign Animal Diseases
FADT WG:	Foreign Animal Disease Threats Working Group
HHS:	Department of Health and Human Services
HSPD:	Homeland Security Presidential Directive
IP:	Intellectual Property
MCM:	Medical Countermeasure
NCAHEM:	National Center for Animal Health Emergency Management
NPV:	Net Present Value
NVS:	National Veterinary Stockpile
PCAST:	President's Council of Advisors on Science and Technology
R&D:	Research and Development
RDT&E:	Research, Development, Test, and Evaluation
RFI:	Request For Information
RFP:	Request For Proposal
US:	United States
USDA:	United States Department of Agriculture

Dr John M. Hardham, Senior Principal Scientist, Global Biologicals Research, Zoetis Inc., 333 Portage Street, KZO-300-203.2, Kalamazoo, MI 49007
Tel: 269-833-0738
Email: john.m.hardham@zoetis.com

Roth JA, Richt JA, Morozov IA (eds): Vaccines and Diagnostics for Transboundary Animal Diseases. Dev Biol (Basel). Basel, Karger, 2013, vol 135, pp 53-58.

Translating Research into Licensed Vaccines and Validated and Licensed Diagnostic Tests

R.E. Hill, Jr., P.L. Foley, N.E. Clough, L.R. Ludemann, D.C. Murtle

USDA, Animal and Plant Health Inspection Service, Veterinary Services, Center for Veterinary Biologics, Ames, IA, USA

Key words: Veterinary; Vaccine; Diagnostic; Emergency; Emerging; Regulatory; Disease; Licensing

Abstract: The USDA Center for Veterinary Biologics (CVB) has the regulatory authority to issue licenses and permits that allow the marketing of pure, safe, potent, and effective veterinary biological products. Under the standard licensing or permitting process, a manufacturer develops, characterizes, and evaluates a product prior to licensure. The CVB evaluates the submitted information, inspects the manufacturing facilities and methods of production and testing, and confirms key product test results through independent testing. This complete and comprehensive evaluation may not be possible during the emergence of a new animal disease or in response to an introduction of a significant transboundary animal disease agent. Processes are in place in the US that allow for more rapid availability of veterinary products in an emerging or emergency animal health situation. But, it can be advantageous to attain preapproval of products prior to their anticipated need. In this article, issues associated with obtaining approval for use of a biological product under emerging or emergency conditions are discussed.

INTRODUCTION

The Virus-Serum-Toxin Act of 1913, as amended in 1985, provides the legal basis for the regulation of veterinary vaccines and related biological products in the United States (US). The Act makes it unlawful to import veterinary biological products into the US, or to ship products anywhere in or from the US, if they are worthless, harmful, contaminated, or dangerous, or if they have not been prepared under a US Department of Agriculture (USDA) license or permit issued by the Center for Veterinary Biologics (CVB). Rules and regulations promulgated to prevent the preparation, sale, barter, exchange or shipment of worthless, harmful, contaminated, or dangerous products are published in Title 9, Code of Federal Regulations (9CFR), Parts 101-118 [1]. Additional program guidance in the form of CVB Notices, Veterinary Services Memoranda, and other guidance documents is available at the CVB website [2].

The CVB is responsible for regulating veterinary biological products, including but not limited to vaccines, bacterins, toxoids, antibodies, and antitoxins intended for use in the treatment of animals and which act primarily through the stimulation, supplementation, enhancement, or modulation of the immune system or the immune response. Because the term "treatment", by definition, includes the prevention, diagnosis, management, or cure of diseases of animals, diagnostic test kits are also under regulatory control by the CVB.

The USDA issues licenses for veterinary biologics produced domestically. In addition to "conventional" or "full" US Veterinary Biological Product Licenses, conditional and autogenous biologics licenses may be issued for specific situations or conditions. The USDA also issues permits to import veterinary biologics into the US. A Permit for Research and Evaluation may be issued if the USDA has sufficient documentation to assess the product and its environmental impact and the product recipient demonstrates the ability to safeguard against any deleterious effects which might result from use of the product. Distribution of the product beyond the original permit recipient is typically prohibited. A Permit for Distribution and Sale is essentially equivalent to a conventional US Veterinary Biological Product License and allows for commercial distribution. For this permit type, the foreign vaccine manufacturer has to coordinate testing and licensure activities with a US company or individual residing within the US, i.e. a permittee who will take responsibility for regulatory compliance with the USDA.

The USDA may specifically exempt any veterinary biological product from one or more of the normal regulatory requirements if this product will be used by, or under the supervision of, the USDA in conjunction with an official USDA program; an emergency animal disease situation; or USDA experimental use of the product. Any licenses and permits issued by the CVB, as well as products exempted by the APHIS Administrator, may include restrictions on the distribution and use of the product. In the event of a foreign disease incursion, for example, distribution may be restricted to USDA disease control activities. Alternatively, distribution in each State may be subject to approval of State Animal Health Officials.

Regulatory Framework for Approving Vaccines and Diagnostic Test Kits in the US

The current USDA regulatory program consists of (1) review of all data generated in support of each product and claim(s) for use; (2) inspection of manufacturing processes and practices including equipment, facilities, materials, personnel, production, quality control, and records; (3) confirmatory testing of biological seeds, cells, and final product by the CVB; (4) a post-licensing monitoring system of inspection and random product testing by the CVB; and (5) post-marketing epidemiological surveillance of product performance under normal conditions of use. This combination of regulatory oversight before and after licensure assures the availability of pure, safe, potent, and effective veterinary biologics to veterinarians and animal owners. Under the standard licensing process, this spectrum of evaluation and review includes the complete characterization, identification, and purity testing of seed materials, product ingredients and the final product; environmental, laboratory, and host animal safety and efficacy studies; stability studies, and post-licensing monitoring of field performance [3,4].

Products must be manufactured from approved Master Seed and Master Cell stocks. The Seeds and Cells are well characterized and tested extensively for purity. Representative samples are then subject to confirmatory testing by the CVB. All product batches (serials) must be derived from an approved Seed or Cell, subject to passage limits.

An "Outline of Production" is required to document how a product is made and must be revised and approved with every change to the process. Manufacturers are subject to on-site inspection prior to licensure, and periodically after licensure, to ensure that facilities are acceptable and procedures described in the Outline are being followed.

All products are evaluated for safety. Monitoring for adverse events following vaccination is an expected feature of all licensing studies. In addition, Seeds for use in live products must be evaluated for reversion to virulence, their ability to be shed by treated animals and infect non-target species, and persistence in the environment. Prior to licensure, each product is also evaluated in a field safety study under normal use conditions, typically conducted in three geographically distinct areas.

Efficacy requirements vary by license type. Each claim of efficacy must be supported, typically by host animal vaccination-challenge data, for full licenses and all permits for distribution and sale. Vaccination effects must be statistically significant and clinically relevant.

Conditional licenses require only a "reasonable expectation" of efficacy. Applicants must provide data to suggest that a product will be efficacious, but product marketing is allowed prior to a definitive demonstration of efficacy. Conditional licenses are intended only as a transitional status and typically expire in one year. Licensees are expected to make continual progress toward full licensure in that time, but conditional licenses may be renewed, if necessary.

Autogenous biologics are custom-made using a Seed (microorganism) isolated from a specific animal herd. They require no Seed approval or efficacy data, but are limited in distribution only to the herd of origin and, under certain circumstances, adjacent herds. All autogenous biologics are inactivated and are prepared by or under the direction of a veterinarian under a veterinarian-client-patient relationship.

The CVB uses risk analysis procedures to evaluate live conventionally-derived vaccines, all biotechnology-derived veterinary biologics, and for proposals to import veterinary biologics into the United States. The CVB has developed Summary Information Formats (SIFs) and Risk Assessment (RA) outlines to facilitate the preparation of scientifically valid and credible risk analyses [5].

Diagnostic test kits are subject to the same requirements for consistent manufacture and adequate facilities as administrable products. Additionally, they must be validated for their intended use. Kits are measured against a defined gold standard assay for diagnostic sensitivity and specificity, as well as dynamic range and analytical sensitivity. Ruggedness, repeatability, and adequacy of the instructions are evaluated in multiple independent laboratories. The intended use, e.g. screening versus confirmatory test; herd status versus individual animal test; differentiation of vaccinated versus infected animals (DIVA) determines specific licensing standards. The advantages of using DIVA vaccines and companion diagnostic test kits for disease eradication and control programs have been presented previously [6].

Considerations for Emerging/Emergency Animal Health Situations

Important factors in deciding whether to use a particular vaccine during an emerging or emergency animal health outbreak include its live or inactivated status, product stability under field conditions, the duration of the immunity provided, the potential development of carriers and masking of infection, the antigenic relatedness of field isolates, and the strain specificity of the vaccine. Important factors in deciding whether to vaccinate at all include feasibility, availability and delivery of vaccine, possible trade implications, and the presence of wildlife hosts or reservoirs. A vaccination strategy (control or eradicate, vaccinate and slaughter) must be selected and an 'exit strategy' developed for use when an outbreak is over [7].

The CVB encourages development and availability of high priority vaccines and diagnostic tests for transboundary animal diseases. The Department of Homeland Security (DHS) and USDA-Plum Island Animal Disease Center (PIADC) are actively developing next-generation vaccines against FMD and other diseases. In addition, veterinary biologics manufacturers are developing new vaccines against emerging and transboundary diseases. These emergency preparedness products may be licensed and distributed under standard processes or, if circumstances require it, with exemption from some or all of the normal regulatory requirements.

Processes used in the US to expedite the availability of veterinary vaccines in an emerging or emergency animal health situation have been previously published [3,4]. They include conditionally licensed vaccines (9 CFR Part 102.6) and autogenous vaccines (9 CFR Part 113.113); experimental (9 CFR Parts 103.3 and 104.4), emergency or official-use vaccine authorizations (9 CFR Part 106.1); and biologics research and evaluation import permits for special distribution of foreign-manufactured biologics that have not completed full USDA approval. The CVB has partnered with other authorities (e.g., the USDA's National Veterinary Stockpile, the Tripartite Canada-Mexico-United States North American Foot and Mouth Disease (FMD) Vaccine Bank) to prepare and purchase vaccine antigen concentrates and finished emergency-use vaccines. Recent examples of products using these processes include experimental use of a new oral rabies vaccine; conditional licenses for a cattle vaccine against FMD and a pandemic H1N1 influenza vaccine for use in swine; and a permit for distribution and sale (restricted by APHIS for as-needed use) for an imported commercial FMD vaccine.

Research Practices that Facilitate Technology Transfer

As a product is developed, the various requirements for regulatory approval should be kept in mind. Researchers should consult with manufacturers and regulators as soon as possible to determine regulatory jurisdiction, obtain comments on a development plan and study protocols, and gain familiarity with the regulations. Research records should be detailed and complete; good laboratory practice (GLP)-like documentation is recommended. Licensing submissions often must be traceable to studies done in a research setting. If no traceability exists, studies may have to be repeated. All data generated in product development must be reported to the CVB.

Master Seeds and Cells must have a defined passage history. Additional documentation is required for seeds that have been genetically engineered; requirements vary according to the type of construct and how it will be used in the final product.

Licensing study results must be traceable to a product made in a specified manner. Every study report should contain an adequate description of product preparation. The format of the Outline of Production (9CFR 114.9) serves as a guideline for the information that should be documented. Documentation of assay validation should be available for any assay methods that may be used in a study.

Prepare for technology transfer to a manufacturing partner as early as possible to avoid concerns about the method of manufacture. Certain licensing studies must be conducted with product prepared by the intended manufacturer in scaled-up production.

If experimental product is to be shipped for testing in animals, advance CVB authorization (9CFR 103.3) and State approval are required, as well as compliance with US Department of Commerce and commercial couriers' shipping regulations. If the product will be used in animals outside of biocontainment, the CVB must receive sufficient information to evaluate risks associated with environmental release. Allow sufficient time to receive shipping authorization prior to anticipated use.

Conclusions

The procedures reviewed above provide a regulatory framework and outline general purity, safety, potency, and efficacy requirements for licensing products for traditional and emerging animal diseases, and the introduction of transboundary animal diseases. The current regulatory system is sufficiently flexible to accommodate a variety of animal health situations while providing the data-driven, performance-based oversight necessary to assure that only quality biologicals are available for use, even in the event of emergencies. The CVB partners with researchers, the biologics industry, and other regulators in proactive vaccine and diagnostic test development. Additional strategies to combat emerging and transboundary animal diseases include vaccine/seed banks or stockpiles, "pre-approved" products to be manufactured upon demand, and knowledge of vaccines and diagnostics available world-wide that may be imported when needed.

References

1 United States Code of Federal Regulations, Title 9: Animals and Animal Products, 2012. [available at: http://www.gpo.gov/fdsys/browse/collectionCfr.action?collectionCode=CFR].

2. USDA Animal and Plant Health, Animal Health. Biologics Regulations and Guidance. [available at: http://www.aphis.usda.gov/animal_health/vet_biologics/vb_regs_and_guidance.shtml].

3. Hill RE, Foley PL, Carr MY, Elsken LA, Gatewood DM, Ludemann LR, Wilbur LA: Regulatory considerations for emergency use of non-USDA licensed vaccines in the United States, in Brown F and Roth J (eds): Vaccines for OIE List A and Emerging Animal Diseases. Dev Biol. Basel, Karger, 2003, vol 114, pp 31-52.

4. Elsken LA, Carr MY, Frana TS, Brake DA, Garland T, Smith K, Foley PL: Regulations for vaccines against emerging infections and agrobioterrorism in the United States of America. Rev Sci Tech Off Int Epiz 2007;26:429-441.

5. USDA Animal and Plant Health, Animal Health. Biologics Regulations and Guidance. Risk Analysis and Summary Information Formats. [available at: http://www.aphis.usda.gov/animal_health/vet_biologics/vb_sifs.shtml].

6. Foley PL, Hill RE: Regulatory considerations for marker vaccines and diagnostic tests in the US. Biologicals 2005;33:253-256.

7. DeHaven WR: Factors to consider when using vaccine to control an exotic disease outbreak, in Brown F and Roth J (eds): Vaccines for OIE List A and Emerging Animal Diseases. Dev Biol. Basel, Karger, 2003, vol 114, pp 281-289.

Richard E. Hill, Jr., USDA, Animal and Plant Health Inspection Service, Veterinary Services, Center for Veterinary Biologics, 1920 Dayton Ave., Ames, Iowa, 50010, USA.

Fax 515 337-7673.

Email: rick.e.hill@aphis.usda.gov

Roth JA, Richt JA, Morozov IA (eds): Vaccines and Diagnostics for Transboundary Animal Diseases. Dev Biol (Basel). Basel, Karger, 2013, vol 135, p 59.

SESSION II

State of the art, progress and gaps in development of vaccines and diagnostics for high priority transboundary diseases for the NVS

Roth JA, Richt JA, Morozov IA (eds): Vaccines and Diagnostics for Transboundary Animal Diseases. Dev Biol (Basel). Basel, Karger, 2013, vol 135, pp 61-72.

Vaccination for the Control of Rift Valley Fever in Enzootic and Epizootic Situations

B. Dungu[1], *M. Donadeu* [1], *M. Bouloy* [2]

1 Global Alliance for Livestock Veterinary Medicines (GALVmed), Edinburgh, Scotland, UK

2 Institut Pasteur, Paris, France

Key words: Rift Valley fever, vaccines, enzootic, epizootic

Abstract: Vaccination continues to be the most effective way to control Rift Valley fever (RVF), a zoonotic insect-borne viral disease of livestock. The irregular, cyclical and persistent nature of RVF in its occurrence in enzootic situations suggests that the vaccination strategy to be considered for these regions should be different from what is envisaged for free from risk regions. Currently available RVF vaccines have been extensively used for the control of the disease. However, these vaccines have shortcomings that have encouraged many research groups to develop new vaccine candidates that would address a large number of the current challenges, and be suitable for use both in disease-free regions and in different contingency and emergency preparedness strategies. The characteristics of different RVF vaccines and vaccination strategies are discussed in this report.

Introduction

Since the first description of a Rift Valley Fever (RVF)-like disease in Kenya in 1912-1913, and the first isolation of the RVF virus in 1931 in Kenya [1], RVF continues to be a major zoonotic and economically important disease of livestock in large regions of Africa and the Middle East, where it is endemic. The disease is also considered to be a big threat to other regions of the world. Countries at risk include those adjacent or close to affected regions (such as the Middle East, Europe and North Africa), and regions further away because the virus now features on most lists of potential biological warfare agents due to its severe zoonotic nature.

RVF is a multi-species insect-borne disease caused by the RVF virus, which belongs to the *Phlebovirus* genus of the *Bunyaviridae* family. The RVF virus is enveloped and spherical with a diameter of 80–120 nm. Like all members of the family, it possesses a single stranded tripartite RNA genome composed of three segments: the large (L), the medium (M), and the small (S) segment. The L segment codes for the RNA-dependent RNA polymerase L protein; the M segment codes for the precursor of the Gn and Gc glycoproteins, and the two non-structural NSm proteins. The S segment codes for both the N nucleoprotein and the non-structural NSs protein, by using an ambisense strategy [2].

A number of factors play a role in the occurrence of outbreaks: (i) the circulation of the RVF virus through mosquito vectors; (ii) the mosquito pressure (number of breeding sites and hatching frequency), which is highly dependent on environmental conditions, particularly rainfall events; and (iii) the distribution of domestic animal hosts, essentially ruminants (goats, sheep and cattle), vulnerable to increased vector/host contacts at night. However, there is extensive species susceptibility to RVF, as set out in Table 1.

Several outbreaks have occurred since the disease was first described in Kenya. Besides those countries linked to the Great Rift Valley formation, which stretch from the Red Sea through East Africa to Madagascar, RVF has occurred in West Africa, with serious outbreaks in Egypt, Mauritania and Senegal. Some of the major outbreaks include:

- 1950-1951: Kenya; 100,000 mortality in sheep; 500,000 abortions [3];
- 1977: Egypt; 200,000 human cases, some 600 reported fatalities [4];
- 1987: Mauritania/Senegal; over 300 human deaths [5];
- 1997-1998: large outbreak in Kenya/Somalia/Tanzania: over 300 human deaths [3];
- 1998-1999: large outbreak in South Mauritania, also in Senegal;
- 2000-2001: appearance of RVF beyond the African region, in Saudi Arabia [6];
- 2007-2008: Sudan, 747 laboratory confirmed human cases, 230 deaths [7];
- 2009-2010: South Africa: 242 lab-confirmed human cases with 26 deaths (unpublished data);
- 2010: Mauritania: 63 human cases, 13 deaths [8];

Several other African countries have also reported small outbreaks or virus circulation as demonstrated by positive serology to RVF antibodies. Although several control measures have been suggested for RVF, vaccination is still the most effective tool, which should be accompanied by other measures such as effective surveillance, good diagnostic strategy and reliable emergency preparedness. This report focuses on the assessment of vaccination strategies for enzootic and epizootic situations.

Table 1: Species susceptibility to RVF [3]

Extremely susceptible (70-100% mortality)	Highly susceptible (20-70% mortality)	Moderately susceptible (less than 10% mortality)	Resistant (inapparent infection)	Refractory (not susceptible)
Lambs	Sheep	Cattle	Camels	Birds
Kids	Calves	Goats	Equids	Reptiles
Puppies		African buffalo	Pigs	Amphibians
Kittens		Asian buffalo	Dogs	
Mice		S. American monkeys	Cats	
Hamsters		Asian monkeys	African monkeys	
Certain other rodents		Certain rodents	Baboons	
		Humans	Rabbits	
			Guinea pigs	
			Certain other rodents	

Current Vaccination Strategies in Enzootic Regions

In regions where outbreaks have occurred, the disease always seems to reoccur. Therefore, eradication or elimination in endemic regions is not a viable option at the moment. Control measures in endemic zones are aimed at early warning, emergency preparedness and minimising the impact of the disease. Measures for the prevention of the disease in regions at risk or free from RVF should be considered differently. In enzootic situations, vaccination is still the most effective method of protecting livestock. Although endemic in most of Sub-Saharan Africa, a very limited number of countries use vaccination for the control of RVF. In countries where vaccination is practiced, different approaches are used, as shown in Table 2.

Characteristics of current RVF vaccines

The advantages and disadvantages of different types of RVF vaccines are set out in Table 3. Two types of vaccines are generally utilised in countries that use vaccination for the control of RVF: live and attenuated vaccines and inactivated vaccines. The most commonly produced and used live RVF vaccines are based on the Smithburn virus derived from a strain isolated from mosquitoes in Western Uganda in 1944 and passaged 79-85 times by intracerebral inoculation of mice: this resulted in the loss of hepatotropism, the acquisition of neurotropism and the capacity to immunise sheep safely when administered parenterally [9]. The South African RVF Smithburn vaccine is based on the 103 mouse brain passage levels of the virus, while Kenya uses the 106 passage level to produce the vaccine, all in Baby Hamster Kidney anchored cell culture systems. The two major producers of the live RVF Smithburn vaccine, Onderstepoort Biological Products (OBP) in South Africa and the Kenya Veterinary Vaccines Production Institute (KEVEVAPI) have

Table 2: Classification of countries based on RVF occurrence and approach to vaccination strategy

RVF Situation	Examples of countries	Current vaccination/Control strategy
Endemic with regular outbreaks	Kenya, Tanzania, Egypt, Madagascar, Sudan	Vaccination at sign of outbreak, Egypt, Sudan: continuous/regular vaccination, No vaccination (Madagascar)
Endemic with sporadic/reoccurring outbreaks	South Africa, Saudi Arabia, Mauritania	Continuous/yearly vaccination , No vaccination (Mauritania)
Endemic with no large scale outbreaks (or not reported), but with serological evidence of virus circulation	Senegal, Mali, DR Congo	No vaccination
Free high risk	Middle East, North Africa	(Active) surveillance
Free low risk	Europe, Americas	Surveillance, talks about vaccine banks

Table 3: Different types of RVF vaccines

VACCINE	STRAIN	ADVANTAGES	DISADVANTAGES
Inactivated, (OBP, VSVRI)	Pathogenic field strain	Safe in pregnant animals, Can be used in outbreak	Short term immunity, Short term immunity, Multiple vaccinations required, Risk of handling virulent strain during production, Colostral immunity present but poor, Sheep better protected than cattle, 100 x more antigen required than for live attenuated, Longer production lead time
Live, Attenuated (OBP, KEVEVAPI)	Smithburn	Highly immunogenic, Single dose, Good immunity (within 21days), Effective and easy production, Safer production, Large batches: >4m doses	Potential residual virulence, Teratogenic for fœtus, Potential risk of reversion to virulence, Not advisable for use in outbreaks, Theoretical possibility of transmission by mosquitoes
Avirulent natural mutant	Clone 13	Good protective immunity in sheep & cattle, Safe in pregnant animals, Safe in outbreak, Produced as standard freeze-dried live vaccine, More than 28 million doses used, Safe, effective and easy to produce, Possible DIVA (NSs ELISA?), Registered & used extensively in South Africa	Only registered to date in South Africa & Namibia, Large scale field data in other regions needed, No evidence of DIVA to date

produced millions of doses since 1952 and 1960 respectively, with the vaccine having been widely used throughout Africa and the Middle East [10].

The live attenuated RVF Smithburn vaccines based on the Smithburn virus have several disadvantages: they may induce abortions, malformations in the foetuses of vaccinated animals, hydrops amnii, and prolonged gestation in a proportion of vaccinated dams. Their use during an outbreak is not advised as they are based on a live virus. Since RVF outbreaks occur in irregular cycles, a number of countries do not to implement vaccination between outbreaks. Another reported problem with the live attenuated RVF Smithburn vaccine is the poor antibody response in vaccinated cattle [11].

In an attempt to address problems associated with the residual virulence of the Smithburn vaccine, as well as the poor antibody response in vaccinated cattle, an inactivated vaccine was developed and has been used since the 1970s in South Africa. This vaccine can be used in all livestock species, at different physiological stages, including pregnancy, and during outbreaks. This inactivated RVF vaccine makes it possible to vaccinate cows that can then confer colostral immunity to their offspring. Given the poor immunogenicity of this vaccine in cattle, it requires a booster three to six months after initial vaccination, followed by annual inoculations [11]. This inactivated RVF vaccine is currently produced in Egypt and in South Africa.

Since 2008, a new vaccine called RVF Clone 13 has been registered and used in South Africa. It is based on an avirulent RVF virus isolated from a non-fatal case of RVF in the Central African Republic that had been passaged in mice and Vero cells, and then plaque purified in order to study the homogeneity of virus subpopulations. A clone designated 13 did not react with specific monoclonal antibodies against NSs and when further investigated was found to be avirulent in mice, yet immunogenic [12]. This vaccine has been evaluated for safety and efficacy in sheep [13] and cattle [14]. More than 10 million RVF Clone 13 vaccine doses were used during the 2009-2010 RVF outbreak in South Africa, and played a key role in the control of the disease (Onderstepoort Biological Products, personal communication).

Considerations for RVF Vaccines

Given the nature of RVF in enzootic regions and the near impossibility of its eradication in these regions, the prevention, control and vaccination strategies to be considered for these enzoonotic regions should be different from those that might be envisaged for free, low and high risk regions. Similarly, the "ideal vaccine" would be different for each of these regions. However, certain generic characteristics should be common to all RVF vaccines and vaccination strategies.

For safety, RVF vaccines should be:

- Safe to produce.- This is not currently the case with the inactivated RVF vaccines, as they are based on a virulent isolate, and are produced in high containment facilities requiring vaccinated personnel.
- Safe for use during all physiological stages of animals.- This is not the case with the current live attenuated Smithburn vaccine which cannot be used in pregnant animals as it might result in abortion and teratogenic effects. Therefore, vaccination strategy is difficult to implement during outbreaks as the pressure to cover all animals is high. In most rural areas of Africa it is also not always possible to determine the pregnancy status of animals, hence many

animals end up being vaccinated while in early pregnancy, which results in low productivity as some early losses of pregnancy are not noticed.

RVF vaccines should present:

- No residual virulence.- This is not the case for the current Smithburn vaccine which may cause abortion in a small proportion of animals, as well as other signs of the disease [15].
- No risk of introduction into the environment.- There is always the risk of the vaccine virus persisting in vaccinated animals and being picked up and transmitted by vectors.
- No risk of reassortment when used in infected animals.- During outbreak there is always a pressure to vaccinate all susceptible animals. It is not always possible to detect incubating animals. An ideal vaccine should not be based on a replicating virus that may reassort with a wild type virus and possibly generate new infectious viruses.
- No risk of spread to humans or other species.- There is almost no data to show whether the Smithburn vaccine, for example, would be safe or not in humans. There is also a risk of the virus establishing itself in other susceptible animals, through mosquito or other vectors.

For efficacy, RFV vaccines should ensure:

- Protection of all susceptible species: preferably the vaccine should trigger a good protective immunity in all susceptible species so that only one type of vaccine is used;
- Quick onset of protective immunity, including in young animals;
- Long lasting immunity;
- Blocking of transmission: the immunity generated by the vaccine should prevent the amplification of RVF virus in ruminants,

RFV vaccines should also:

- Be cost effective for producers and users.
- Require only a single vaccination: This is not the case with the current inactivated RVF vaccine. It would also be preferable if the vaccine could trigger a lifelong immunity.
- Be easy to apply of application.
- Be suitable for stockpiling (vaccine or antigen bank) and quick availability.

Considerations for Vaccination Strategies for Enzootic Regions

Regions where RVF is enzootic could consider the following issues:

- Vaccination frequency: A yearly vaccination and regular vaccination of weaners is recommended and practiced in some countries. There is a widespread practice of considering vaccination at the first indication of an outbreak or of when it can be predicted. However, the reality is that it is difficult to predict an outbreak more than three months in advance; and this is not enough to carry out an effective vaccination programme. This has been confirmed by studies of outbreaks in Kenya, Tanzania, Sudan, etc. [16]. Hence the need has arisen to consider regular or yearly vaccination.

- Improvement of herd immunity: Given the cyclical nature of RVF outbreaks in enzootic situation, vaccination is often not conducted in a sustained way during inter-epidemic periods. One way of contributing to improved uptake during these periods would be to use multivalent vaccines that include RVF with a vaccine already used regularly. Regular vaccination with a multivalent vaccine would increase RVF herd immunity. One such multivalent vaccine is being developed to include a capripox vaccine, the lumpyskin virus vaccine, and the RVF Clone 13 vaccine. The advantage of the lumpyskin virus vaccine is that it can be used for the respective cattle disease, as well as for sheep and goat pox diseases in these small ruminants.
- Use of sentinel animals: It has been a practice in some enzootic countries to position naive sentinel animals in high risk areas and monitor them for seroconversion as an early warning mechanism [17]. Unfortunately, this approach has been difficult to enforce in many countries due to the complex logistical requirements such as the need for regular sampling and testing. In most cases these sentinel animals are located far from laboratories with the required diagnostic capacity. Sustained monitoring is difficult because of the high cost of existing tests.

Other attributes that would be desirable for the vaccine to be used in these regions include thermotolerance and small pack sizes. Limited infrastructures, high ambient temperatures and the need to cover sometimes long distances on poorly maintained or inexistent roads often complicate the maintenance of the vaccine at recommended temperatures, even when cooler bags or portable fridges are used. A vaccine that would be less labile to high ambient temperatures for a number of hours would therefore be preferable. Given the fact that some livestock units in village settings of the affected regions have small number of animals, reconstituting standard freeze-dried vaccine vial of 100 or 200 doses tends to lead to wastage or forces vaccinators to maintain the reconstituted vaccine for longer than the recommended time, thus compromising its potency.

Vaccination Strategy for Regions Free of RVF

For regions free of RVF, where the disease would have been introduced for the first time, the decision to vaccinate requires a thorough analysis of the circumstances. In case vaccination has to be considered, the characteristics of the RVF vaccine to be used would preferably be different from those of the vaccine used in enzootic situations. Considerations on a vaccination strategy for epizootics in previously free regions should include not only the generic characteristics described above, but also:

- Protection for young animals and possibly newborn naïve animals: Young ruminants are the most susceptible group of animals, and in endemic regions they are more likely to have maternal immunity, which would not be the case with naive dams.
- Sterilizing immunity: This would be preferable in order to stop the transmission of the virus and prevent its spread into new reservoirs.
- Availability of a vaccine and companion diagnostic test that allow differentiation between infected and vaccinated animals (DIVA). This might consist of a positive and/or negative marker vaccine, useful for export of animals from infected countries, and for detecting infection, respectively.
- The use of a non-replicating or replication deficient RVF vaccine: This would be preferable to minimise the risk of persistence of any form of vaccine antigen in animals or the environment.

For a region where the infection has already been introduced, it would be important to eliminate possible sources of re-infection. This should include a culling strategy and movement control.

RVF Vaccine Development

To date, in addition to the three RVF vaccines currently registered and in use (i.e. the live attenuated Smithburn and Clone 13 vaccines, and the inactivated RVF vaccine), several strategies have been employed to develop a wide range of vaccines; and candidate vaccines are at different stages of development [18]. Most of these technologies were reviewed during the 2011 RVF meeting organised by the FAO [18]. Table 4 summarises some characteristics of different candidate vaccines evaluated to date.

The development and evaluation of new RVF vaccine candidates will always be challenging because of a number of factors, including the zoonotic nature of the virus and the disease, the lack of a widely available human vaccine and the difficulty in reproducing consistently the disease in target animals. Key challenges in developing new RVF vaccines include the following needs:

(i) Biosecurity Level (BSL) 3 to 4 stables for animal work, with laboratory capacity (serology, virus isolation, virus titration);

(ii) Vaccinated personnel if animal challenge studies with a virulent virus have to be conducted, unless it is in a very high containment facility that would allow working on animals without direct contact; and

(iii) An adequate and proven challenge model: Most efficacy studies in recent years have relied on demonstrating the ability of the vaccine to prevent abortion in vaccinated pregnant ewes challenged with a virulent RVF virus [13, 19]. This has been justified by the difficulty in consistently reproducing the clinical disease or major signs of the disease in experimental infection. The use of the above challenge model requires the involvement of well-trained veterinarians and adequate equipment and facility to conduct oestrus synchronization and subsequent artificial insemination of the dams. However, through recent work by Kortekaas et al. [20], challenge models based on the ability of the vaccine to reduce or prevent viremia in vaccinated animals post-challenge can now be used to evaluate the efficacy of a RVF vaccine.

Conclusion: toward Effective RVF Control Strategies

One of the biggest challenges to the control of RVF through the use of vaccines remains the design of suitable vaccination strategies. The cyclical nature of the disease and the subsequent long inter-epidemic periods have resulted in many countries considering the use of vaccination only at first indication of an outbreak. Control strategies that promote the establishment of good herd immunity in the livestock population in endemic regions are more likely to reduce the devastating impact of severe outbreaks. As discussed earlier, a number of factors prevent many endemic countries from putting into place effective control strategies for RVF. Key obstacles to effective RVF control include:

- The cyclical nature of RVF and long inter-epidemic period;
- The specific policies on the control of RVF or lack of emergency preparedness strategy for RVF outbreaks in many countries;
- Poor diagnostic capacity when the diagnostic laboratory is located far from high risk areas where the disease tends to start, while limited resources prevent the conduct of RVF activities outside outbreak seasons; and
- The high pressure on the few RVF vaccine manufacturers, because of their limited production capabilities, when there are outbreaks in more than one country.

In order to address these challenges and work toward more effective emergency preparedness, the Global Alliance for Livestock Veterinary Medicines (GALVmed) has been involved in a number of initiatives in three endemic regions of Africa, i.e. Southern, Western and East Africa. These initiatives are additional to the work aimed at supporting the development of appropriate control tools that address the need for improved RVF control especially in poor livestock keepers' communities. The control tools currently under development include:

- A combination RVF and Lumpyskin disease vaccine which would rely on the regular vaccination of animals for Lumpyskin disease in cattle and sheep and goat pox in sheep and goat respectively, in order to build immunity in vaccinated animals to RVF;
- A RVF field test in the form of lateral flow device or pen-side test, which allows for testing animals in situ, avoiding costly shipment of serum specimens to laboratories; and
- A RVF vaccine/vaccine antigen bank for a monovalent RVF vaccine.

Table 4: Candidate RVF vaccines evaluated in target animals

VACCINE	STRAIN	ADVANTAGES	DISADVANTAGES
Live attenuated	MP12	Effective and good protective immunity, Easy and safe to produce, Better safety than Smithburn in most species and age groups, Dual vaccine, Safe in all animals, DIVA, Long shelf life (LSD), More thermo-tolerant than others, Efficacy shown in animal trials	Teratogenic for fœtus, Not available commercially
Recombinant Lumpy skin virus expressing RVF	LSD Neethling strain expressing RVF glycoproteins	Dual vaccine Safe in all animals DIVA Long shelf life (LSD) More thermo-tolerant than others Efficacy shown in animal trials	Only proof of concept to date, Currently grown in primary cells, Possible GMO regulation challenge (?)
Recombinant-multiple deletion virus	Reverse genetic generating RVF virus with double deletions in NSs & NSm [16]	Less prone to reassortment, Live vaccine, DIVA: negative marker, Easy and safe to produce	No published proof of concept in target animals
Virus-vectored RVF vaccines	Heterologous virus, Newcastle disease virus expressing RVF GP [17]	DIVA: Positive & Negative marker, Live vaccine, Replication deficient, Multivalent: suitable where annual vaccination is a challenge, Potential for improved thermostability	No registered vaccine yet available, No large scale field data yet available, although extensive analytical data generated
Avirulent (lab generated) reassortant	R566: deletion in the M and S segments	Safer due to deletions in all 3 segments, may never reassort, Protection in mice	No field evaluation to date
Virus like particle (VLP)	VLP made of envelop proteins (GP) [18]	Potentially very safe, Immunity similar to live vaccine, but no replication, DIVA	No proof of concept in target animals, Large scale production might be a challenge
DNA	DNA priming + inact. Vaccine, [19],cDNA encoding GP, [20]	DIVA, Potentially long lasting immunity, Ability to enhance and modulate induced immunity	Only incomplete protection demonstrated in mice, Production challenges, Regulatory challenges (use in food animals)

Regional strategies for the control of RVF are expected to encourage more cohesive policies and mutual support between different countries. A number of initiatives are already under way in the Southern Africa Development Community (SADC) region. These initiatives include the evaluation of RVF risk in each country through the development of risk maps based on historic data on the occurrence of the disease [3], a RVF policy landscaping in order to understand current policies setup in each country around RVF control, and the establishment of a technical RVF interstate working group that would advise policy makers on the most appropriate RVF control strategy, which would include aspects such as vaccination strategy and support for the establishment of a common vaccine or vaccine antigen bank.

In conclusion, effective vaccines and vaccination strategies have been developed, and great progress has been achieved in understanding the pathogenesis of the disease and developing emergency preparedness and control strategies to RVF. However, there are still a number of issues that require further attention and research. These include:

- The need for a human RVF vaccine;
- More effective field diagnostic tools;
- Effective surveillance strategies;
- Better prediction tools for endemic regions;
- Better understanding of inter-epidemic periods, and possible evolution or changes in virus virulence: what make some outbreaks in endemic areas more severe than others, or affect specific species more severely? (the last SA outbreak was severe in cattle, which had not been the case before);
- Development of effective strategies that will contribute to maintaining herd immunity, thereby preventing serious outbreaks; and
- Strengthening the local and regional capacity in RVF control, which would require confronting the disease, both in livestock and humans.

AckowledgEments

The authors would like to thank Onderstepoort Biological Products for supplying information on the RVF Clone 13 vaccine and the Southern African Development economic community's RVF Working group for sharing their regional RVF control strategy.

References

1. Daubney R, Hudson JR,Garnham PC: Enzootic hepatitis or Rift Valley fever: An undescribed disease of sheep, cattle and man from East Africa. J Path Bact 1931; 34: 545-579.
2. Giorgi C, Accardi L, Nicoletti L, Gro MC, Takehara K, Hilditch C et al.: Sequences and coding strategies of the S RNAs of Toscana and Rift Valley fever viruses compared to those of Punta Toro, Sicilian Sandfly fever, and Uukuniemi viruses. Virology 1991;180:738–753.
3. Murithi RM, Munyua P, Ithondeka PM, Macharia JM, Hightower A, Luman ET, et al: Rift Valley fever in Kenya: history of epizootics and identification of vulnerable districts. Epidemiol Infect 2011;139 (3):372- 380.
4. Meegan JM, Hoogstraal H, Moussa MI: An epizootic of Rift Valley fever in Egypt in 1977. Vet Rec 1979;105(6):124-125.
5. Zeller HG, Fontenille D, Traore-Lamizana M, Thiongane Y, Digoutte JP: Enzootic activity of Rift Valley fever virus in Senegal. Am J Trop Med Hyg 1997;56(3):265–272.
6. Madani TA, Al-Mazrou YY, Al-Jeffri MH, Mishkhas AA, Al-Rabeah AM, Turkistani AM: Rift Valley fever epidemic in Saudi Arabia: epidemiological, clinical, and laboratory characteristics. Clin Infect Dis 2003;37:1084–1092.
7. Hassan OA, Ahlm C, Sang R, Evander M: The 2007 Rift Valley fever Outbreak in Sudan. PLoS Negl Trop Dis 2011;5(9):e1229.

8. El Mamy AB, Baba MO, Barry Y, Isselmou K, Dia ML, El Kory MO, H, Diallo M, El Kory M, Diop M, Lo M, Thiongane Y, Bengoumi M, Puech L, Plee L, Claes F, de La Rocque S and Doumbia B: Unexpected Rift Valley fever outbreak, northern Mauritania. Emerg Infect Dis 2011;17(10):1894-1896.
9. Smithburn KC: Rift Valley fever: the neurotropic adaptation of the virus and the experimental use of this modified virus as a vaccine. Br J Exp Pathol 1949;30:1-16.
10. Swanepoel R, Coetzer JAW. Rift Valley fever. In: Coetzer JAW &Tustin RC (eds). Infectious diseases of livestock; 2nd ed. South Africa: Oxford, Oxford University Press, 2004; 1037–1070.
11. Barnard BJH: Rift Valley fever vaccine: antibody and immune response in cattle to a live and inactivated vaccine. J. S. Afr. vet. Assoc. 1979; 50:155-157.
12. Muller R, Saluzzo JF, Lopez N, Dreier T, Turrel M, Smith J, et al.: Characterization of Clone 13, a naturally attenuated avirulent isolate of RVF virus, which is altered in the small segment; Am J Trop Med Hyg 1995;53:405– 411.
13. Dungu B, Louw I, Lubisi A, Hunter P, von Teichman BF, Bouloy M: Evaluation of the efficacy and safety of the Rift Valley Fever Clone 13 vaccine in sheep. Vaccine 2010;28 (June (29)):4581–4587.
14. von Teichman B, Engelbrecht A, Zulu G, Dungu B, Pardini A, Bouloy M: Safety and efficacy of Rift Valley fever Smithburn and Clone 13 vaccines in calves. Vaccine 2011;29 (August (34)):5771–5777.
15. Botros B, Omar A, Elian K, Mohamed G, Soliman A, Salib A, Salman D, Saad M, Earhart K: Adverse response of non-indigenous cattle of European breeds to live attenuated Smithburn Rift Valley fever vaccine. J Med Virol. 2006 Jun;78(6):787-91.
16. Anyamba A, Linthicum KJ, Small J, Britch SC, Pak E, de La Rocque S, Formenty P, Hightower AW, Breiman RF, Chretien JP, Tucker CJ, Schnabel D, Sang R, Haagsma K, Latham M, Lewandowski HB, Magdi SO, Mohamed MA, Nguku PM, Reynes JM, Swanepoel R: Prediction, assessment of the Rift Valley fever activity in East and Southern Africa 2006-2008 and possible vector control strategies. Am J Trop Med Hyg. 2010 Aug;83 (2 Suppl):43-51.
17. Clements AC, Pfeiffer DU, Martin V, Pittliglio C, Best N, Thiongane Y: Spatial risk assessment of Rift Valley fever in Senegal; Vector Borne Zoonotic Dis. 2007 Summer;7(2):203-16.
18. FAO: Rift Valley fever vaccine development, progress and constraints. Proceedings of the GF-TADs meeting, January 2011, Rome, Italy. FAO Animal Production and Health Proceedings, No. 12. Rome, Italy
19. Hunter P, Erasmus BJ Vorster JH: Teratogenicity of a mutagenised Rift Valley fever strain (MVP12) in sheep. Onderstepoort J. vet. Res. 2002;69 (1): 95-98.
20. Kortekaas J, Antonis AF, Kant J, Vloet RP, Vogel A, Oreshkova N, de Boer SM, Bosch BJ, Moormann RJ: Efficacy of three candidate Rift Valley fever vaccines in sheep. Vaccine 2012;14;30 (23):3423-9. Epub 2012 Mar 24
21. Bird BH, Albariño CG, Hartman AL, Erickson BR, Ksiazek TG, Nichol ST Rift valley fever virus lacking the NSs and NSm genes is highly attenuated, confers protective immunity from virulent virus challenge, and allows for differential identification of infected and vaccinated animals. J Virol. 2008 Mar;82 (6):2681-91. Epub 2008 Jan 16.
22. Kortekaas J, Dekker A, de Boer SM, Weerdmeester K, Vloet RP, de Wit AA, Peeters BP, Moormann RJ Intramuscular inoculation of calves with an experimental Newcastle disease virus-based vector vaccine elicits neutralizing antibodies against Rift Valley fever virus. Vaccine 2010; 28 (March 11):2271-6. Epub 2010 Jan 14.
23. Näslund J, Lagerqvist N, Habjan M, Lundkvist A, Evander M, Ahlm C, Weber F, Bucht G: Vaccination with virus-like particles protects mice from lethal infection of Rift Valley Fever Virus. Virology. 2009 Mar 15;385 (2):409-415. Epub 2009 Jan 20.
24. Lorenzo G., Martín-Folgar R., Rodríguez F., Brun A.: Priming with DNA plasmids encoding the nucleocapsid protein and glycoprotein precursors from Rift Valley fever virus accelerates the immune responses induced by an attenuated vaccine in sheep; Vaccine 2008, 26, 5255-5262
25. Lagerqvist N, Näslund J, Lundkvist A, Bouloy M, Ahlm C, Bucht G: Characterisation of immune responses and protective efficacy in mice after immunisation with Rift Valley Fever virus cDNA constructs. Virol J. 2009; 6 (Jan (6))

Baptiste Dungu, GALVmed, Edinburgh, EH26 0PZ Scotland, United Kingdom
Tel: +44 0131 445 6198
Email: Baptiste.dungu@galvmed.org;

Roth JA, Richt JA, Morozov IA (eds): Vaccines and Diagnostics for Transboundary Animal Diseases. Dev Biol (Basel). Basel, Karger, 2013, vol 135, pp 73-78.

Diagnostic Approaches for Rift Valley Fever

W.C. Wilson [1], H.M. Weingartl [2], B.S. Drolet [1], K. Davé [3], M.H. Harpster [4], P.A. Johnson [4], B. Faburay [5], M.G. Ruder [1], J.A. Richt [5], D.S. McVey [1]

1 Arthropod-Borne Animal Diseases Research Unit (ABADRU), Agricultural Research Service (ARS), US Department of Agriculture, Manhattan, KS, USA

2 National Centre for Foreign Animal Disease (NCFAD), Canadian Food Inspection Agency (CFIA), Winnipeg, Manitoba, Canada

3 VecTOR Test Systems, Inc., Thousand Oaks, CA, USA

4 Department of Chemical and Petroleum Engineering, University of Wyoming, Laramie, WY, USA

5 Department of Diagnostic Medicine and Pathobiology, Kansas State University, Manhattan, KS, USA

Key words: Rift Valley fever, diagnosis, detection

Abstract: Disease outbreaks caused by arthropod-borne animal viruses (arboviruses) resulting in significant livestock and economic losses world-wide appear to be increasing. Rift Valley fever (RVF) virus is an important arbovirus that causes lethal disease in cattle, camels, sheep and goats in Sub-Saharan Africa. There is concern that this virus could spread because of global warming, increased animal trade or through bioterrorism. This paper discusses the current and developing approaches to diagnosis of RVF. Diagnostic assays are available for RVF, but availability can be limited and there is a need for global harmonization. Continued improvement of standard serological and viral genome amplification approaches, including new embedded/syndromic testing, biosensor, emerging virus detection and characterization technologies is needed.

INTRODUCTION

Rift Valley fever (RVF) outbreaks since 2006 in Kenya, Tanzania, Sudan, Mauritania, Madagascar and South Africa have heightened awareness of the potential spread and impact of this viral disease. High mortality rates in young small and large ruminants and high abortion rates are associated with RVF virus (RVFV) outbreaks. Global warming has been hypothesized as a reason for the recent increased reporting of arboviral diseases. Whether true or not, the potential for RVFV to be introduced into non-endemic countries through increased animal trade or as a potential bioterrorism agent is widely recognized. The intentional or unintentional introduction of RVFV into non-endemic countries is of agricultural and public health concern. RVFV is a Category A pathogen and a high priority for biodefense. The disease is enzootic in sub-Saharan Africa with RVF epidemics occurring at irregular intervals primarily in eastern and southern Africa, generally associated with periods of above average rainfall. In 2000, however, cases of RVF were confirmed in both humans and livestock in Saudi Arabia and Yemen, marking the first occurrence of the disease outside of the African continent. Therefore, there is a need for safe, reliable and affordable diagnostic assays that can be used in endemic countries where RVF is a real animal and public health issue, as well as tools for early detection of an introduction into naive countries.

Handling of diagnostic specimens and using diagnostic assays is complicated by the fact that RVFV is zoonotic and infection often results in mild to lethal disease in humans. RVFV can also result in retinal degeneration leading to permanent blindness in 10% of infected humans. Vaccination does provide added safety when working with the pathogen; however, there are a very limited number of veterinary diagnosticians that are immunized with the expensive investigational human vaccine available only from USAMRIID. Therefore, rapid, sensitive, specific methods that quickly inactivate the virus during sample processing and allow sensitive and specific diagnosis are important. In addition, diagnostic assays which are compatible with the new vaccines are needed that ensure the ability to differentiate infected from vaccinated animals (DIVA). Finally, multiplex assays and other tools capable of early detection and characterization of emergent viruses are essential for the rapid identification of newly introduced viruses.

SEROLOGICAL ASSAYS (ELISA, IHC)

Classical viral diagnostic assays such as virus isolation (VI) and virus neutralization (VN) tests pose health risks to laboratory staff, especially with the limited availability of the human experimental vaccine and vaccinated people. VN can be performed with the Smithburn or MP-12 vaccine strains to reduce risks but there are still risks from handling the sample. Furthermore, the VI and VN tests require proper facilities and personal protective equipment (PPE) that are not always available. Most laboratories use enzyme-linked immunosorbent assays (ELISAs) that have been developed to detect antibodies to RVFV in humans, sheep and cattle. Although antigen capture ELISAs have been developed [1], they are not commonly used. Safer to produce and distribute are ELISAs based on recombinant nucleocapsid protein (Np) [2]. Although some of these assays are commercially available, they are very expensive and not always immediately available during times of high need. Therefore, we have also developed indirect and competitive ELISAs based on recombinant Np; and to increase specificity we have developed ELISAs based on recombinant Gn and Gc.

To develop an assay compatible with the RVF NSs deleted vaccine and other vaccine candidates, we have been evaluating indirect ELISAs based on recombinant NSs. In experimental RVFV infections, animals developed anti-NSs antibodies (either IgM or IgG) later than virus neutralizing antibodies; and the anti-NSs IgM antibodies were detected in a limited number of animals. However, anti-NSs IgG antibodies were detected in all animals at 28 days post infection (dpi). In contrast, virus neutralizing antibodies were present in all serum samples at 7 dpi until the end of the experiments. Dot blots and western blots performed on these sera did not detect anti-NSs antibodies at 7 dpi or earlier. In summary, indirect ELISAs and other serological methods detected anti-NSs antibodies in fewer serum samples than compared with the plaque reduction neutralization assay; therefore, RVFV NSs may not be a consistent DIVA marker early after infection. The competitive ELISA based on RVF NSs may be more sensitive but has yet to be fully evaluated.

Lateral flow assays for RVF based on expressed antigens are also in development by several research groups including ours. Preliminary evaluations of this simple, straight forward and field deployable technology are very promising. This system could provide a thermostable pathogen-specific tool in areas where diagnostic tools are not readily available to local or regional veterinary diagnostic laboratories. The various ELISAs and lateral flow assays are currently undergoing further development to improve their robustness. Once established, these assays will require international cooperation to achieve full evaluation and validation.

Tissues are usually the first samples received by diagnostic laboratories during RVF outbreaks. However, receiving fresh or frozen tissues in good enough condition for virus isolation is rare. Additionally, aborted tissue and tissue fluids become heavily contaminated with environmental bacteria and yeast, which interfere with PCR diagnostics. Because of this, and the safety hazards posed by fresh tissues during RVF outbreaks, tissues should be fixed in formalin in the field before submission to diagnostic labs. Once fixed, virus is inactivated and tissues are safe to handle for routine histopathological evaluation and immunohistochemical (IHC) detection of viral antigen. To date, however, there are no commercial BSL-2 anti-RVFV reagents available for IHC by veterinary diagnostic laboratories in non-endemic countries. Antibodies that do exist for IHC are typically produced in BSL-3 or BSL-4 biocontainment laboratories against whole wild-type virus and therefore must be used in endemic countries or in high biocontainment laboratories. Recently, an IHC assay was developed using a rabbit anti-nucleocapsid protein antibody which can be used safely in BSL-2 diagnostic laboratories [3]. The assay performs equally well on infected cell cultures, insect and mammalian tissues.

Genomic Amplification Assays

Another method that allows rapid inactivation of the RVFV in clinical samples, yet allows early detection of this deadly and zoonotic virus, is genomic amplification technology. The most common is reverse transcriptase–polymerase chain reaction (RT-PCR). Standard RT-PCR assays are available for both mammalian and mosquito samples [2]. Closed tube real-time RT-PCR methods are favored, because they are less labor intensive and less prone to contamination. Three real-time RT-PCR assays have been developed for RVFV [4-7]. These assays all target different segments of the tripartite genome; and two of the assays were shown to be effective with clinical samples [6, 7]. We have applied these assays to samples generated during experimental infections of sheep, cattle and mosquitoes with both virulent and vaccine RVFV strains. Several commercial extraction kits were evaluated. The extraction method of choice is primarily based on throughput and approved institutional safety protocols. Previously, we have described the adaptation of a real-time RT-PCR assay [6] to a rapid inactivation extraction followed by a single-step real-time RT-PCR [3]. We have also generated a more robust single tube multiplex real-time RT-PCR assay that detects all three genome segments. This assay uses different reporter dyes for each segment and an external RNA amplification control using a fourth reporter dye. This format provides diagnostic confirmation and is DIVA compatible in that the assay can distinguish between viral RNA from wild-type versus viral RNA from an NSs gene deleted vaccine. This real-time RT-PCR assay has been evaluated in endemic countries (unpublished data). In addition, there are two real-time reverse transcription-loop-mediated isothermal amplification (LAMP) assays available that provide alternative amplification detection tools for RVF viral RNA [8, 9].

New Technologies (Embedded Biosensor, Emerging Disease Detection)

In the event of a RVFV introduction into a naïve country, local and regional veterinary diagnostic laboratories would lack the capacity to test for RVFV. This is why the concept of embedded diagnostic testing for the rapid detection and characterization of emergent viruses is gaining support. Several tools have been reported such as GeneChip technology, MassTag PCR, and high-throughput deep sequencing [10].

The pan-pathogen detection tools have primarily been developed for human pathogens, especially respiratory pathogens [11], although there have also been applications for veterinary disease agents [12]. There is at least one commercial human disease gene array available [11]. A less expensive, higher throughput technology is MassTag PCR that has been successfully applied to the detection of infectious diseases including arthropod-borne viruses [12]. High-throughput sequencing or re-sequencing approaches have also proven to be a viable pathogen detection and characterization tool [10, 12]. Most of these assays are nucleic acid based. There are developing technologies for antigen and serum antibody detection assays such as liquid microarray [13], peptide arrays [14] and biosensor tools [15]. Biosensor technology has also been applied to nucleic acid detection [16]. The range of new technologies to facilitate rapid detection, syndromic or multiplex diagnostic assays and pathogen characterization are expanding

Conclusions

Standard RVFV diagnostic tools are available but pose health risks to laboratory personnel especially in areas where proper facilities and PPE are limited or not available. The most common assays for RVF diagnosis are ELISAs, real-time RT-PCR and LAMP assays. Although a comparison of the current ELISA systems has been reported recently [17], there is limited standardization and availability of reagents. To increase detection of RVFV introduction into a naïve country, embedded testing similar to the multiplex vesicular disease panel [18] is needed. Diagnostic technology is a rapidly advancing and evolving area, and it is unclear what technology or combination of technologies will be the most useful and widely adopted by diagnostic laboratories. However, the development of harmonized diagnostic tools is needed for effective detection of emergent or exotic viruses such as RVFV.

Acknowledgements

The funding for the project was provided from USDA-ARS CRIS project number 5410-32000-005-00D, Canadian CRBNE Research and Technology Initiative project number CRTI-06-0138RD, the U. S. Department of Homeland Security Grant Award Number 2010-ST-AG0001 to CEEZAD, The Kansas Bioscience Authority Trust Agreement Number 54-5430-1-403 and through an interagency agreement with the Science and Technology Directorate of the U.S. Department of Homeland Security under Award Number HSHQDC-07-00982.

The contents of this publication are solely the responsibility of the authors and do not necessarily represent the official views of CFIA, KSU, DHS, KBA or the USDA.

REFERENCES

1 Morvan J, Saluzzo JF, Fontenille D, Rollin PE, Coulanges P: Rift Valley fever on the east coast of Madagascar. Res Virol 1991;142:475-482.

2 Anonymous: Rift Valley Fever, in OIE Terrestrial Manual. 2008, pp 323-333.

3 Drolet BS, Weingartl HM, Jiang J, Neufeld J, Marszal P, Lindsay R, et al: Development and evaluation of one-step rRT-PCR and immunohistochemical methods for detection of Rift Valley fever virus in biosafety level 2 diagnostic laboratories. J Virol Methods 2012;179:373-382.

4 Garcia S, Crance JM, Billecocq A, Peinnequin A, Jouan A, Bouloy M, Garin D: Quantitative real-time PCR detection of Rift Valley fever virus and its application to evaluation of antiviral compounds. J Clin Microbiol 2001;39:4456-4461.

5 Drosten C, Gottig S, Schilling S, Asper M, Panning M, Schmitz H, Gunther S: Rapid detection and quantification of RNA of Ebola and Marburg viruses, Lassa virus, Crimean-Congo hemorrhagic fever virus, Rift Valley fever virus, dengue virus, and yellow fever virus by real-time reverse transcription-PCR. J Clin Microbiol 2002;40:2323-2330.

6 Bird BH, Bawiec DA, Ksiazek TG, Shoemaker TR, Nichol ST: Highly sensitive and broadly reactive quantitative reverse transcription-PCR assay for high-throughput detection of Rift Valley fever virus. J Clin Microbiol 2007;45:3506-3513.

7 Njenga MK, Paweska J, Wanjala R, Rao CY, Weiner M, Omballa V, et al: Using a field quantitative real-time PCR test to rapidly identify highly viremic Rift Valley fever cases. J Clin Microbiol 2009;47:1166-1171.

8 Le Roux CA, Kubo T, Grobbelaar AA, van Vuren PJ, Weyer J, Nel LH, et al: Development and evaluation of a real-time reverse transcription-loop-mediated isothermal amplification assay for rapid detection of Rift Valley fever virus in clinical specimens. J Clin Microbiol 2009;47:645-651.

9 Peyrefitte CN, Boubis L, Coudrier D, Bouloy M, Grandadam M, Tolou HJ, Plumet S: Real-time reverse-transcription loop-mediated isothermal amplification for rapid detection of rift valley Fever virus. J Clin Microbiol 2008;46:3653-3659.

10 Lipkin W: Microbe hunting. Microbiol Mol Biol Rev 2010;74:363-377.

11 Chen EC, Miller SA, DeRisi JL, Chiu CY: Using a pan-viral microarray assay (Virochip) to screen clinical samples for viral pathogens. J Vis Exp 2011;(50). pii: 2536. doi: 10.3791/2536.

12 Palacio G, Briese T, Lipkin WI: Microbe hunting in laboratory animal research. ILAR J 2010;51:245-254.

13 Perkins J, Clavijo A, Ortiz JI, Salo TJ, Holland HJ, Hindson BJ, McBride MT: Toward a multiplexed serotyping immunoassay for foot-and-mouth disease virus. J Vet Diagn Invest 2007;19:180-184.

14 Duburcq X, Olivier C, Malingue F, Desmet R, Bouzidi A, Zhou F, et al: Peptide-protein microarrays for the simultaneous detection of pathogen infections. Bioconjug Chem 2004;15:307-316.

15 Neng J, Harpster MH, Wilson WC, Johnson PA: Surface-enhanced Raman scattering (SERS) detection of multiple viral antigens using magnetic capture of SERS-active nanoparticles. Biosens Bioelectron 2012;26(3):1009-1015.

16 Zhang H, Harpster MH, Wilson WC, Johnson PA: Surface-enhanced Raman scattering detection of DNAs derived from virus genomes using Au-coated paramagnetic nanoparticles. Langmuir 2012;28:4030-4037.

17 van Vuren PJ, Paweska JT: Comparison of enzyme-linked immunosorbent assay-based techniques for the detection of antibody to Rift Valley fever virus in thermochemically inactivated sheep sera. Vector Borne Zoonotic Dis 2010;10:697-699.

18 Lenhoff RJ, Naraghi-Arani P, Thissen JB, Olivas J, Carillo AC, Chinn C, et al: Multiplexed molecular assay for rapid exclusion of foot-and-mouth disease. J Virol Methods 2008;153:61-69.

Dr William C. Wilson, Research Microbiologist, Arthropod-Borne Animal Diseases Research Unit, Center for Grain and Animal Health Research, USDA, Agricultural Research Service, 1515 College Ave., Manhattan, KS 66502, USA.

Tel: (785) 537-5570 Fax: (785)-537-5560

Email: william.wilson@ars.usda.gov

Roth JA, Richt JA, Morozov IA (eds): Vaccines and Diagnostics for Transboundary Animal Diseases. Dev Biol (Basel). Basel, Karger, 2013, vol 135, pp 79-94.

Current Status and Future Needs in Diagnostics and Vaccines for High Pathogenicity Avian Influenza

D.E. Swayne, E. Spackman

Exotic and Emerging Avian Viral Diseases Research Unit, Southeast Poultry Research Laboratory, Agricultural Research Service, US Department of Agriculture, Athens, GA, USA

Key words: Avian influenza, diagnostics, vaccination, vaccines

Abstract: Since 1959, 32 epizootics of high pathogenicity avian influenza (HPAI) have occurred in birds. Rapid detection and accurate identification of HPAI has been critical to controlling such epizootics in poultry. Specific paradigms for the detection and diagnosis of avian influenza virus (AIV) in poultry vary somewhat among different countries and industry compartments depending on specific needs and resources. Importantly, since HPAI and low pathogenicity (LP) AI of the H5 and H7 subtypes are reportable to the World Organization for Animal Health (OIE), diagnostic procedures are implemented for regulatory purposes and are harmonized to some degree. Most current tests are adequate and have been in use for some time, therefore they have been well validated and presently there is no reported new technology that will completely replace the current tests. However, some modifications, updates or additional tests could be beneficial. The element of AIV diagnostics that is most in need of improvement is in determining the hemagglutinin and neuraminidase subtype specificity of antibody to AIV. Most HPAI epizootics have been eradicated using traditional stamping-out programs, but beginning in 1995, five epizootics have added vaccination as an additional, interim control tool. From 2002-2010, >113 billion doses of AI vaccine have been used in poultry; 95.5% as oil-emulsified, inactivated whole AIV vaccines and 4.5% as live vectored vaccines. The majority of vaccine has been used in the four H5N1 HPAI enzootic countries (China [91%], Egypt [4.7%], Indonesia [2.3%], and Vietnam [1.4%]) where vaccination programs are directed to all poultry. The 10 other countries/regions have used less than 1% of the vaccine, administered in a focused, risk-based approach. Some vaccine "failures" have resulted from antigenic drift of field viruses away from the vaccine viruses, but most have resulted from failures in the vaccination process; i.e. failure to adequately administer the vaccine to at risk poultry resulting in lack of population immunity. China, as the major AIV vaccine user, will drive innovation and commercialization of new vaccine technologies, but because of the low-cost to manufacture the current high quality inactivated whole AIV vaccines, such vaccines will continue to dominate the market for the next 10 years.

Introduction

Since 1955, when the etiology of fowl plague was determined to be influenza A virus, 32 epizootics of high pathogenicity avian influenza (HPAI) have occurred in avian species, mostly domestic poultry [1]. Historically, HPAI virus (HPAIV) has caused sporadic epizootics in domestic poultry when a low pathogenicity (LP) avian influenza virus (AIV) is introduced from the wild bird reservoir and allowed to circulate in poultry, mutating from low to high virulence [2]. However, since 1996, the H5N1 HPAIV has become enzootic in poultry within several countries which has necessitated two main changes within HPAI control and eradication strategies; 1) development and implementation of rapid diagnostic tests to accelerate diagnosis before the virus spreads, which permits a quicker stamping-out action leading to eradication, and 2) addition of vaccines and vaccination as a control tool to manage clinical disease, prevent human infections and maintain food security, especially in economically disadvantaged countries.

Diagnostics

Background and state of diagnostic science

Although specific diagnostic paradigms and work flow may vary depending on needs, the same tests are used almost universally for avian influenza virus (AIV) detection, antibody detection, and subtype identification of AIV from poultry (Fig. 1). Specific procedures for each of these tests are recommended by the World Organization for Animal Health (OIE) (ww.oie.int), and Food and Agriculture Organization Animal Influenza (OFFLU) network, which afford some level of harmonization. All of the current procedures have been in place for many years.

The speed with which HPAI is eradicated is greatly dependent upon the identification of the index case and elimination of the disease before it spreads to additional premises. This was not possible using the traditional diagnostic tool of virus isolation, and necessitated a broad clinical case definition along with quarantine and depopulation of poultry within 0.5-3 km zone in order to compensate for delayed diagnosis. However, with the advent of molecular diagnostics, specifically the real-time reverse transcriptase polymerase chain reaction (rRT-PCR) assay, a diagnosis of AI could be made within 3 hours of sample submission. The AIV rRT-PCR test was developed in 2000-2001 by USDA, and field tested in the H7N2 LPAI live poultry market depopulation program during December 2001 in New York and New Jersey [3] and, again during the winter/spring 2002 in Virginia commercial poultry industries [4]. The first field use of rRT-PCR against HPAI was in Chile (H7N3) during late spring of 2002, and in Hong Kong (H5N1) during September 2002 (D. Suarez, personal communication, 8-20-2012). A conventional reverse transcriptase polymerase chain reaction (cRT-PCR) assay was used during the 2003 Dutch H7N7 HPAI outbreak to allow rapid decisions for control [5].

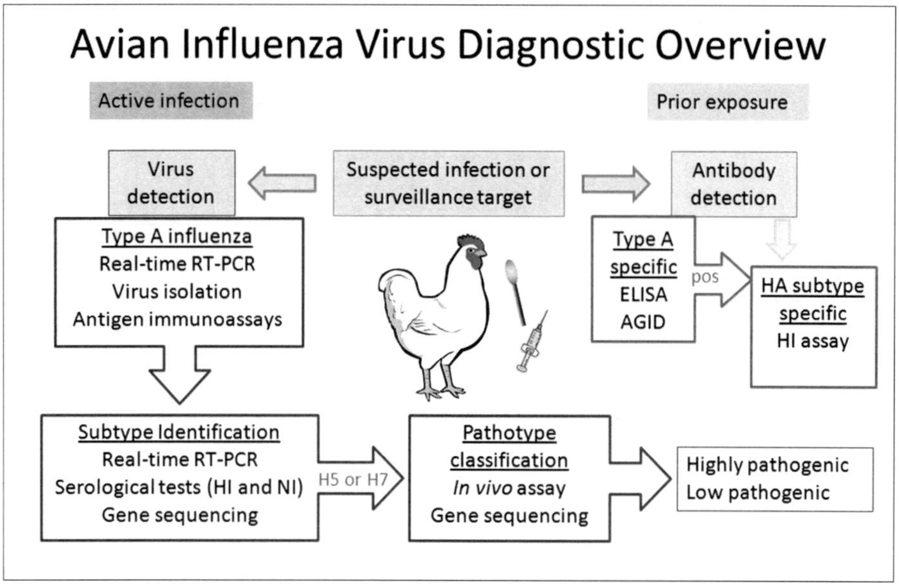

Fig.1. Cartoon of the diagnostic flow for avian influenza virus (AIV) and antibody to AIV. Virus detection tests are conducted with oral swabs, cloacal swabs or tissues. The 3 primary virus detection tests are virus isolation, reverse transcription polymerase chain reaction (RT-PCR) and antigen immune-assays. If a specimen is positive with any of these tests, the subtype may be identified with serological tests, RT-PCR or gene sequencing. If the isolate is determined to be an H5 or H7 subtype virus, the pathotype is characterized by in vivo testing and gene sequencing. Antibody assays to evaluate prior exposure may be type specific (i.e. for type A influenza), such as enzyme linked immunosorbent assay (ELISA) or agar gel immune-diffusion (AGID), or may be subtype specific (often for H5 or H7) such as hemagglutination inhibition (HI) assay.

The strengths and weaknesses of each diagnostic method are well known and each has been periodically modified to achieve the best results, which includes balancing the logistical workflow as well as sensitivity and specificity. In addition to the OIE and OFFLU recommended tests, which are public domain, commercial kits are available for rRT-PCR, antigen detection and ELISA.

The characteristics of the perfect diagnostic assay are: fast, cheap, simple to conduct/easy to use, robust, sensitive and specific. No current test meets all of these criteria. Numerous novel methods and technologies, such as new instruments, (e.g. biosensors [6-9]), new approaches (e.g. microarray [10]) and modified versions of current tests (e.g. microsphere based serology [11-13]) have been reported, but none have been validated, which is partly because none have offered enough of an advantage over the current methods to justify the time and expense of validation. Most of the recent work in AIV diagnostics has been to refine current technology like molecular methods which have a continual need to be updated to account for sequence changes in the virus in the field. Initial validations of some have been reported [14-17].

Type A specific tests

Virus isolation in embryonating chicken eggs (ECE) and rRT-PCR (or other RT-PCR based methods) are presently used to detect type A influenza virus directly from field specimens. Rapid type specific tests, rRT-PCR and antigen immunoassays (AgIA), may also be used as screening tests prior to processing a specimen for virus isolation. Virus isolation will unlikely be completely replaced since an isolate is necessary for diagnostic confirmation and characterization.

Although current type specific methods for detecting AIV (i.e. methods for detecting influenza A virus) are adequately sensitive and specific, in most cases, there are areas where improvement or additional tools would be beneficial. For example, more efficient and rapid methods for RNA extraction from tissues and cloacal swabs that remove inhibitors and stabilize the RNA would aid rRT-PCR. Virus isolation would benefit from a media (or other method) which could stabilize viable virus in the absence of the cold chain.

Antigen detection methods such as the commercial AgIA are helpful in the field for inexpensive, rapid testing and can be used pen-side, but they lack sensitivity. Improved sensitivity would greatly improve the utility of these tests. Additionally, a Newcastle disease virus (NDV) AgIA test would help with rapid differentiation of AIV and NDV in the field.

Subtype identification

Field specimens are frequently tested for the H5, H7 or other relevant hemagglutinin (HA) subtypes directly by rRT-PCR. Once an isolation is made, the subtype is identified by serology (hemagglutination inhibition (HI) assay and neuraminidase inhibition (NI) assay) and/or by sequencing the HA and NA. Because HI and NI are resource intensive and large reagents sets are needed to identify all 16 HA and 9 NA subtypes, serological subtyping is most often performed only in reference laboratories. In other labs, serological subtyping may focus on only the most critical HA subtypes, H5 and H7, or other relevant subtypes for the region. Sequencing takes longer (up to 3 days), and not all labs are equipped to sequence specimens as the instrumentation is expensive and requires advanced technical skills to use, but sequencing is the most accurate.

Although not widely used, there are H5 specific AgIA kits available commercially. However, most are not validated adequately to demonstrate that they have sufficient sensitivity with all antigenic variants. Rapid H5 and H7 subtype tests, such as AgIA, which can be validated for specificity would be beneficial for rapid diagnosis in the field or initial testing of egg fluid from an isolate.

Pathotype identification

Once the H5 or H7 subtype has been identified in a specimen, the pathotype must be determined. Pathotype classifications are defined by the OIE, therefore the tests are prescribed by the OIE. Currently there are two prescribed tests: in vivo chicken testing and determining the sequence of the HA proteolytic cleavage site

(PCS). Due to the resources needed for each of these tests, and the regulatory nature of pathotype classification, testing is typically only conducted in national or regional reference laboratories. In vivo chicken testing is the most biologically relevant, but it is very resource intensive and the full test takes up to 10 days after the virus has been isolated in a culture system. Sequencing is much faster and may be accomplished with just the RNA from a swab or tissue specimen, however as discussed previously the resources for sequencing are not universally available.

A rapid test for pathotype classification would be very beneficial, but no technology or method had been identified which would fulfill this purpose. RT-PCR based methods have been proposed and reported [18], but have not been validated and it is not clear if it would be accurate with the wide sequence variation among the PCS of both HP and LP isolates. Furthermore, sequence outside of the PCS may affect pathotype due to post-translational modifications (i.e. glycosylation)[19], so any tests which only assay the PCS may not be accurate with all isolates.

Serology

There are two assays routinely used for pan-AIV antibody detection in poultry, ELISA (numerous commercial kits are available worldwide) and agar gel immunodiffusion (AGID) assay. Both are type specific, relatively inexpensive and have similar sensitivity. Because of the rapid speed and simplicity of high throughput, ELISA will likely eventually replace AGID. Additionally, unlike AGID which does not work well with sera from waterfowl, some commercial ELISA kits are not species specific so they can be used with sera from domestic ducks and other poultry, or even wild bird species. On the contrary AGID is cheaper on a per sample basis and since it detects IgM instead of IgY, antibodies can be detected earlier after infection, so the AGID does have some advantages in certain situations. However, the positive AGID tests results drop-off after 3 months as IgM AIV-specific antibody production is switched to IgY, but ELISA can detect antibody beyond this time period.

Probably the element of AIV diagnostics where there is the most need for innovation is for methods to characterize the subtype specificity of antibodies to AIV. Cross reaction among subtypes and inhibition from the NA can make HI assay results difficult to interpret or even inaccurate. A few solutions have been proposed which involve producing the reference antigens for HI assay in vitro using systems which only express the HA and not the NA protein. However, these would be expensive to develop, produce and validate. Other alternatives are using expressed or whole virus proteins with other serological test platforms, such as microsphere assays, which would allow a greater number of antigens to be screened with a single specimen, thus could provide more in depth and possibly accurate data.

Vaccines and Vaccination

Background

Historically, eradication has been the desired outcome and has been achieved through comprehensive, integrated control programs utilizing: 1) education of farmers and governmental officials in disease control (including behavioral change communications), 2) diagnostics and surveillance to identify infected flocks, 3) enhanced biosecurity (including quarantine of infected flocks, movement controls within the outbreak zone, and cleaning and disinfection of premises and equipment), and 4) elimination of infected poultry through culling or depopulation [1]. However, in 1994, an epizootic of H5N2 HPAI occurred in central Mexico overwhelming the resources of the poultry industries and government necessitating the implementation of a fifth tool, i.e. vaccination, which allowed management of the clinical disease for eventual eradication. Vaccination has since been implemented in Pakistan against H7N3 HPAI during late 1995; 13 Asian, European and African countries against H5N1 HPAI from 2002 to present; North Korean against H7N7 HPAI during 2005; and Mexico against H7N3 HPAI beginning in 2012 [20, 21]. Vaccination has emerged to be a valuable tool for interim management of HPAI, assuring national food security and preserving the livelihood of rural poor, especially in least developed, developing and transition countries [20, 22]. Vaccination did not create the enzootic nature of H5N1 HPAI in China, Egypt, Indonesia or Vietnam; enzootic status was already evident in these countries prior to implementing vaccination. However, routine and improper use of vaccines and vaccination has contributed to prolonging enzootic status of H5N1 HPAI, contributing to virus "persistence" in the field by complicating surveillance and contributing to complacency.

Vaccines in the field

Based on a survey for 2002-2010, over 113 billion doses of AIV vaccine have been used in poultry in 15 different countries/special administrative regions, mainly against the H5N1 HPAI panzootic within Asia, Africa and Europe. China has been the main user with >103 billion doses (90.99%), followed by Egypt (5 billion doses; 4.65%), Indonesia (2.6 billion doses, 2.32%), and Vietnam (1.6 billion doses, 1.43%), with all 4 countries having enzootic status for H5N1 HPAI and using the vaccines in nationwide routine vaccination programs for all poultry [20]. This high number of vaccine doses used in China was linked to correspondingly high poultry populations in China as compared to a much lower populations in Egypt, Indonesia, and Vietnam. The other 11 countries/special administrative regions (Hong Kong, Mongolia, Kazakhstan, France, The Netherlands, Cote d'Ivoire, Sudan, North Korea, Israel, Russia, and Pakistan) have used <1%, and such vaccine use has been limited to preventative or emergency vaccination programs in limited geographic areas or specific poultry species of higher risk, and not in nationwide vaccination programs. In mid-2012, Mexico began using AIV vaccine in response to an outbreak of H7N3 HPAI in the center of the country [21].

Of the 113 billion doses of vaccine used, 95.5% has been inactivated whole AIV vaccines, while 4.5% has been live recombinant virus vaccines [20]. The latter was primarily recombinant NDV (rNDV) vectored vaccine with H5 influenza gene insert, but some recombinant fowlpox virus (rFPV) with H5 or H5 plus N1 gene inserts has been used. Most recently, a recombinant herpesvirus turkey (rHVT) with an H5 gene insert and recombinant duck virus enteritis virus (rDVE) with an H5 gene insert has been developed, laboratory tested and licensed or a license is pending [23]. A rFPV with an H5 gene insert has been used in Central America since 1998 against H5N2 LPAI and over 2 billion doses have been used through 2005 [24].

Despite the broad use of large quantities of AIV vaccines against H5N1 HPAI since 2002, vaccine use has not resulted in eradication in the four enzootic countries; i.e. China, Egypt, Indonesia and Vietnam. In addition, there have been reports of AIV vaccine "failures", specifically isolation of H5N1 and/or identification of clinical disease consistent with HPAI in vaccinated flocks or in regions that vaccinate against H5N1 reporting resurgence of HPAI disease cases [22]. Some of these "failures" have been the result of vaccines not protecting against field challenge because of antigenic drift of the field viruses away from the vaccine strains, but more often, the "failures" have resulted from the lack of proper administration of vaccine or inability to vaccinate and produce a protective immune response in > 60-80% of the at risk poultry population; i.e. a failure to achieve population immunity because of the inability to vaccinate poultry properly [25].

Vaccine technology

The vaccine technologies used experimentally and licensed for use in poultry in the field are listed in Table 1. Five different categories of vaccines have been examined in the laboratory as discovery projects or proof of principle: inactivated whole AIV, live AIV, live vector, in vitro produced hemagglutinin, and DNA vaccines. However, licensing and use has only been achieved with a small number of the technologies and products: inactivated whole AIV vaccines (95.5% of vaccine used) and live vectored vaccines (4.5%; avian paramyxovirus type 1/rNDV, rFPV, rHVT and rDVE). When contemplating why more new technologies have not made it into the market place, the new technology should be compared against a list of the ideal traits of AIV vaccines and against practical knowledge of how poultry health problems are resolved (Table 2). In addition, the reader must accept that the ideal vaccine for humans may not be ideal for poultry.

Before any new technologies will be adopted in poultry, the new vaccine must be shown to protect in experimental poultry trials and must provide protection that is equivalent to or better than the "gold standard", oil-emulsified whole AIV vaccine. In commercial poultry, vaccine development and use is driven by economics and thus adoption of new technologies occurs only if it provides an economic advantage that costs less than living with the problem, or partial solution provided by existing products. Historically, the majority of AIV vaccines have been inactivated whole AIV which are produced in embryonating chicken eggs. This mature, established technology has been successfully used for over 40 years to produce trillions of doses of potent and efficacious inactivated or live attenuated vaccines to other poultry diseases such as Newcastle disease, infectious bronchitis, infectious bursal disease, and other viral pathogens. This technology is inexpensive and has produced

Table 1. Experimental and licensed vaccines for high pathogenicity avian influenza in poultry and other avian species (reviewed in [50].

Vaccine Category	Vaccine	Species	Route	AI Subtypes	HPAIV Challenge Tested	Licensed	Comments	Additional Ref.
Inactivated AIV	Adjuvanted whole avian influenza virus	Chicken (layer & broiler), turkey, duck, geese, other poultry, zoo birds	SQ, IM, in ovo	H5, H7	Yes	Yes	Mostly oil-emulsified; some with aluminum hydroxide. Includes H5 and H7 LPAIV, H5 and H7 HPAIV and H7 reverse genetic LPAIV seed strains. Requires parenteral administration	
Live AIV	Live wild-type LPAI virus	Chicken	IM, IT Spray	H5, H7	Yes	No	Rumors of intentional exposure with LPAIV to protection from HPAIV have been reported in H5N1 and H5N2 HPAI outbreaks in 1990's and 2000's	[32, 33]
	Attenuated LPAI virus	Chicken	Spray	H5, H7	Yes	No	Temperature sensitive mutant or replace HA with ectodomain of NDV HN gene; risk assessment needed for reassortment potential	[34, 35]
Live Vector	rd-Adenovirus	Chicken	SQ, IN, In ovo	H5	Yes	No	rd = Replication defective, only 1 round of replication occurs after injection. SQ and *in ovo* protected	[36]
	Avian leukosis virus	Chicken	IM	H7	Yes	No		[33]
	Avian paramyxovirus type 1 (rNDV)	Chicken	Eye, IN	H5, H7	Yes	Yes	Licensed in Mexico and China	[37, 38]
	Duck enteritis virus (rDVE)	Duck	IM	H5	Yes	No	Submitted for license in mid-2012 for ducks (China)	[23]
	Fowlpox virus (rFPV)	Chicken (layer & broiler), goose, Muscovy ducks	SQ, WW	H5, N1	Yes	Yes	Licensed 1997 for chickens (USA, Mexico); used primarily in Central America against H5N2 LPAI; limited use in China	
	Herpesvirus Turkey (rHVT)	Chickens	SQ	H5, N1	Yes	Yes	Licensed 2012 for chickens (USA, Egypt). Used Egypt	[39]
	Infectious laryngotracheiti s virus vector	Chicken	Eye	H7, H5, N1	Yes	No	N1 did not protect	[40, 41]
	att-Salmonella typhimurium	Chicken	Oral	H5, M2 e	Yes	No	Attenuated vaccine strain. Failed to protect from HPAIV challenge with single oral immunization	[42, 43]
	Vaccinia	Chicken	IM, IP	H5	Yes	No	Low to no antibody response	[44]
	rd-Venezuelan Equine Encephalitis virus		SQ, *in ovo*	H5	Yes	No	rd = Replication defective, only 1 round of replication occurs after injection	[45]
In vitro produced hemagglutinin	Baculovirus in insect cell culture	Chicken, Muscovy duck	SQ	H5, H7	Yes	No		[46]
	Eukaryotic systems (plants or cells cultures)	Chicken	IM	H5	Yes	No	*Nicotiana* sp.	[47]
DNA	Naked DNA	Chicken	IM	H5	Yes	No	Not financially viable. Improvements needed in promotors and adjuvants to decrease quantity of nucleic acid needed and reduce number doses for protection	[48, 49]

Table 2. Properties of ideal avian influenza vaccines and vaccination methods for poultry.

Desired Property	Current Situation
Inexpensive	Current cost for inactivated AIV vaccine: $0.05-0.10/dose plus cost of administration ($0.05-0.07 per dose for individual handling and injection) [50]
Use in multiple avian species	Most used in meat, layer and breeder chickens, but large quantity also used in ducks; minor amounts in turkeys, geese, quail, etc. [20]
Single dose protection	Most situations require minimum of 2 doses; prime-boost scenario is optimal with additional boost in long-lived birds at 6-12 month intervals [51, 52]
Mass application	95.5% is inactivated vaccine administered by handling and injecting individual birds, with 4.5% as vectored vaccine given by mass spray vaccination (rNDV vector) [20]
Identify infected birds in vaccinated population	Serological differentiation tests are available, but only minor use. Most vaccine applied without using a DIVA strategy [52]
Overcome maternal antibody block	Maternal antibody to AIV hemagglutinin or virus vector inhibits primary immune response. Initial vaccination must be timed for declining maternal antibody titers to allow optimal primary immune response [53], as also is needed decline in active immunity before giving booster vaccinations [54]
Given at 1 day of age in hatchery or *in ovo*	Inactivated vaccine provide poor protection if given at 1 day of age. Vectored vaccines can be given at 1 day of age, but generally require a field boost with inactivated vaccine 10 days or more later
Antigenically close to field virus	The majority of inactivated whole AIV vaccine uses reverse genetic generated vaccine seed strains to antigenically match field viruses [20, 25]

effective, potent vaccines, and lacks the additional cost of royalty payments for patents used and the expense to purchase and use new manufacturing equipment which would accompany implementation of newer vaccine technologies. However, newer technologies will and have been adopted at the higher cost when they addressed one or more traits that make the new vaccine closer to the ideal vaccine (Table 2). For example, the majority of inactivated vaccine doses produced and used in poultry during 2007-2012 have utilized reverse genetic technology to generated AIV vaccine seed strains which are closer antigenically to H5N1 viruses encountered in the field, and these vaccine seed strains have been updated when antigenic variants have developed in the field [25].

Because of the economic aspect of poultry vaccine adoption and long-withdrawal period for oil adjuvant, use of inactivated whole AIV vaccines has been limited to valuable, long-lived or specialty poultry within developed countries, but in developing/transitional countries, the lower cost of labor for vaccination and shorter withdrawal periods for oil adjuvants has fueled use of vaccines in the much larger populations of meat chickens and ducks. Experimental studies have demonstrated the possibility for mechanized in ovo injection for oil-emulsified, inactivated whole AIV vaccines [26], adenovirus vectored vaccines [27], VEE-vectored vaccine [28], attenuated AIV vaccine [29], and rNDV vectored [30] vaccines for low-cost mass administration, but this has not been commercially implemented. However, theoretically, an even better approach would be to develop vaccine technologies for applying vaccines via low cost methods such as spray application for respiratory delivery or per os application for oropharyngeal and upper digestive tract delivery (i.e. feed or water).

However, even with such technological breakthroughs, the next fundamental question is whether vaccination to control HPAI is the best approach, or if other control strategies such as prevention through management biosecurity or, if encountering a HPAI epizootic, if immediate stamping-out is the better approach. In a recent survey, most countries preferred using a stamping-out program without vaccination and indicated vaccination would only be used if the HPAI epizootic was large and stamping-out was not effective at immediate eradication or vaccination could be used as a potential preventative measure for valuable poultry when the threat of an epizootic was high [20]. Historically, this threshold of overwhelming HPAI epizootic and implementation of vaccination has been reached more quickly in least developed and developing/transition countries than in developed countries, as evident within the 15 countries that vaccinated of which 13 were least developed or developing/transition, but only two were developed countries (France and The Netherlands) and the latter two used a small, time-limited targeted vaccination program [31, 20].

Vaccination

Even with use of vaccine seed strains that antigenically match field viruses and with commercial vaccines of high potency, protection in the field can only be achieved if the at risk poultry are immunocompetent and if individual birds receive the vaccine in the proper dose (which may include booster vaccinations) and by the correct administration route. The goal is to achieve population immunity in the at risk poultry population, which is only achieved if >60-80% of the poultry have a protective immune response [20]. From a national perspective using routine vaccination of all poultry, this goal of national population immunity is difficult to achieve because of limited financial and human resources. Of the 113 billion doses of AIV vaccine used 2002-2010, this resulted in only a 41.9% coverage rate among the at-risk national poultry population of the 15 vaccinating countries/regions [20]. Of these 15, five conducted routine vaccination campaigns of all poultry. Based on their individual poultry populations and vaccination protocols, coverage rate for China was 47.1%, Hong Kong was 86.2%, Egypt was 69.9%, Indonesia was 14% and Vietnam was 52.3% [20]. This suggest that only two countries/regions achieved population immunity, i.e. Hong Kong and Egypt, but alternative data which presented higher village poultry population estimates for Egypt suggests the vaccine coverage rate in Egypt was substantially lower, between 27.8-48.6%

[20]. In addition, the use of 1-day-of-age vaccination in broilers in Egypt may have also contributed to poor population immunity. H5N1 HPAI cases in poultry continue in China, Egypt, Indonesia and Vietnam because of the lack of population immunity, but Hong Kong did achieve national population immunity, with only one H5N1 HPAI case in poultry between 2003-2012 [22]. These data suggest that achieving national population immunity in poultry, with its intensive financial and human resource requirements, is not feasible in most countries. Instead, vaccination should be targeted to at risk poultry and/or specific geographic regions, which requires field surveillance and epidemiological data and modeling to design effective vaccination programs. In addition, the historical yearly vaccination campaigns in cattle and pigs for transboundary diseases are not effective with poultry because of the shorter generation times (i.e. chickens and ducks have 5 month generation time), allowing production of large naïve poultry populations between the vaccination campaigns, thus perpetuating the disease. In addition, countries with large semi-commercial and village poultry populations must develop unique programs that will reach the large number of households with low numbers of birds. Initially, expectations were high that a spray vaccination of rNDV would provide single dose, uniform protection in all poultry. Experimentally, rNDV-H5-AIV by respiratory mass application did provide protection from HPAIV challenge in the laboratory, but when deployed to the field, the presence of high levels of maternal antibody to NDV and H5 AIV, inhibited rNDV replication and failed to provide protection with the single vaccine dose. Instead, the rNDV-H5-AIV has best been used as a priming vaccine followed by inactivated booster vaccination. Additional research is need on optimizing vaccination protocols for different poultry species and ages to achieve low cost immunity.

Conclusions

In the review of available literature and field data the following are concluding observations and specific future needs concerning AIV diagnostics and vaccines for poultry over the next 10 years:

- Current diagnostic methods for AIV in poultry are tied to regulatory needs. Virus detection, subtype identification and pathotyping tests are prescribed by the OIE and any changes will need to fulfill regulatory needs.
- Although regulatory considerations harmonize AIV diagnostics, specifics of testing varies among countries and industry compartments to satisfy different needs and to accommodate the availability of resources.
- In order to be implemented new test formats and modifications, tests must:
  - Undergo thorough validation
  - Be practical for the field and veterinary diagnostic laboratory workflow
  - Be cost effective

- Elements of AIV diagnostic programs where improvements are needed (important, but are generally not critical)
  - A method or media to stabilize viable virus for transport in the absence of the cold chain
  - A rapid and high throughput RNA extraction method for tissues and cloacal swabs which removes substances that are inhibitory for rRT-PCR and which preserves the RNA
  - An AgIA for Newcastle disease virus as differential test for AIV, and improved AgIA's for type A influenza which have sensitivity which is near to that of virus isolation or rRT-PCR (currently the difference is 103 to 105 50% egg infectious doses of virus)
  - A sensitive, specific, and well validated AgIA for identification of the H5 and H7 HA subtypes which reacts with all known variants within each subtype.
  - A rapid and inexpensive test to determine pathotype in a biologically relevant manner
  - A method to accurately identify the HA specificity of antibody to AIV (type A influenza). This is a critical need
- Current test are adequate for most AIV testing, however improvements and modifications would be beneficial
- China and other developing/transition countries will be the primary users and will drive commercialization of new technologies for vaccines, with China being the major user, driving innovation and commercialization of vaccine technologies for domestic demand and a modest export market to other developing/transition countries
- Inactivated whole AIV vaccines will dominate the market, but antigenically-relevant, reverse genetic generated AIV seed strains will continue to be developed and used as field viruses drift antigenically away from current AIV vaccine seed strains, and usage of such seed strains will become more focused to specific national/regional geographic needs as H5N1 HPAI became more geographically-isolated and continue to evolve into different genetic subclades or antigenic subgroups
- Developed countries will focus more on preventative measures such as enhanced biosecurity and early detection, with stamping-out of infected and epidemiologically linked premises, while vaccines and vaccination will only have minor usage as preventative tools used in focused, risk-based strategies
- The current routine nationwide HPAI vaccination programs should be converted to risk-based programs based upon real-time field surveillance with vaccine use being focused to the poultry species, region, and production sectors that are at highest risk of HPAI outbreaks and reservoirs; and these vaccination programs should be continuous and age-specific, and not based on seasonal campaign system, designed to achieve 60-80% coverage in at risk poultry populations and provide continual feedback to modify the vaccination program

as local conditions change. DIVA strategies should be incorporated for improved epidemiological surveillance and vaccination systems

- Improved vaccination rates are needed among domestic ducks, which are a large asymptomatic reservoir of H5N1 HPAI virus in Asia. Vaccination of ducks should be linked to control of an economically-significant disease such as duck virus enteritis (DVE), potentially using a recombinant live DVE vaccine with H5 gene insert, thus providing protection from both DVE and H5 HPAI in a single vaccine.
- Additional research and control program needs:
  - New vaccine vector platforms are needed for cost-effective vaccination of short-lived meat chickens which does not require catching and handling of individual birds, possibly achieved by spray, per os or in ovo application, and ideally providing protection after a single vaccination and not inhibited by maternal immunity
  - A mass vaccination system is needed for village poultry with possible development of a thermostable vaccine that can be used within existing NDV village vaccination programs to boost participation and coverage rates
  - More prime-boost vaccination protocols need to be developed to maximize protection at least cost and provide better population immunity
  - Newer adjuvants for inactivated vaccines are needed that will provide shorter withdrawal periods and that can elicit stronger, longer-lived mucosal immune responses
  - A more time-responsive vaccine licensing and registration process is needed by national veterinary biological authorities; for example, the development of a process of updating replaceable "vaccine cassettes" without requirement of resubmitting a full licensing dossier.

References

1. Swayne DE, Suarez DL, Sims LD: Influenza, in Swayne DE, Glisson JR, McDougald LR, Nair V, Nolan LK, Suarez DL (eds): Diseases of Poultry. Ames, Iowa, Wiley-Blackwell, 2013, in press.
2. Rohm C, Horimoto T, Kawaoka Y, Suss J, Webster RG: Do hemagglutinin genes of highly pathogenic avian influenza viruses constitute unique phylogenetic lineages? Virology 1995;209:664-670.
3. Spackman E, Senne DA, Myers TJ, Bulaga LL, Garber LP, Perdue ML, et al: Development of a real-time reverse transcriptase PCR assay for type A influenza virus and the avian H5 and H7 hemagglutinin subtypes. J Clin Microbiol 2002;40:3256-3260.
4. Akey BL: Low pathogenicity H7N2 avian influenza outbreak in Virginia during 2002. Avian Dis 2003;47:1099-1103.
5. Elbers ARW, Fabri THF, de Vries TS, de Wit JJ, Pijpers A, Koch G: The highly pathogenic avian influenza A (H7N7) virus epidemic in The Netherlands in 2003--lessons learned from the first five outbreaks. Avian Dis 2004;48:691-705.
6. Lai WA, Lin CH, Yang YS, Lu MS: Ultrasensitive and label-free detection of pathogenic avian influenza DNA by using CMOS impedimetric sensors. Biosens Bioelectron 2012;35(1):456-460.

7. Wang R, Lin J, Lassiter K, Srinivasan B, Lin L, Lu H, et al: Evaluation study of a portable impedance biosensor for detection of avian influenza virus. J Virol Methods 2011;178(1-2):52-58.
8. Lum J, Wang R, Lassiter K, Srinivasan B, Abi-Ghanem D, Berghman L, et al: Rapid detection of avian influenza H5N1 virus using impedance measurement of immuno-reaction coupled with RBC amplification. Biosens Bioelectron 2012;38:67-73.
9. Xu J, Suarez D, Gottfried DS: Detection of avian influenza virus using an interferometric biosensor. Anal Bioanal Chem 2007;389:1193-1199.
10. Metzgar D, Myers CA, Russell KL, Faix D, Blair PJ, Brown J, et al: Single assay for simultaneous detection and differential identification of human and avian influenza virus types, subtypes, and emergent variants. PLoS One 2010;5:e8995.
11. Cha W, Ma Y, Saif YM, Lee CW: Development of microsphere-based multiplex branched DNA assay for detection and differentiation of avian influenza virus strains. J Clin Microbiol 2010;48:2575-2577.
12. Lupiani B, Mozisek B, Mason PW, Lamichhane C, Reddy SM: Simultaneous detection of avian influenza virus NP and H5 antibodies in chicken sera using a fluorescence microsphere immunoassay. Avian Dis 2010;54:668-672.
13. Kuriakose T, Hilt DA, Jackwood MW: Detection of avian influenza viruses and differentiation of H5, H7, N1, and N2 subtypes using a multiplex microsphere assay. Avian Dis 2012;56(1):90-96.
14. Pedersen J, Killian ML, Hines N, Senne D, Panigrahy B, Ip HS, Spackman E: Validation of a real-time reverse transcriptase-PCR assay for the detection of H7 avian influenza virus. Avian Dis 2010;54:639-643.
15. Slomka MJ, Pavlidis T, Coward VJ, Voermans J, Koch G, Hanna A, et al: Validated RealTime reverse transcriptase PCR methods for the diagnosis and pathotyping of Eurasian H7 avian influenza viruses. Influ Other Respir Viruses 2009;3:151-164.
16. Slomka MJ, Densham ALE, Coward VJ, Essen S, Brookes SM, Irvine RM, et al: Real time reverse transcription (RRT)-polymerase chain reaction (PCR) methods for detection of pandemic (H1N1) 2009 influenza virus and European swine influenza A virus infections in pigs. Influ Other Respir Viruses 2010;4:277-293.
17. Spackman E, Ip HS, Suarez DL, Slemons RD, Stallknecht DE: Analytical validation of a real-time reverse transcription polymerase chain reaction test for Pan-American lineage H7 subtype Avian influenza viruses. J Vet Diagn Invest 2008;20(5):612-616.
18. Leijon M, Ullman K, Thyselius S, Zohari S, Pedersen JC, Hanna A, et al: Rapid PCR-based molecular pathotyping of H5 and H7 avian influenza viruses. J Clin Microbiol 2011;49(11):3860-3873.
19. Ohuchi M, Orlich M, Ohuchi R, Simpson BEJ, Garten W, Klenk HD, Rott R: Mutations at the cleavage site of the hemagglutinin alter pathogenicity of influenza virus A/chick/Penn/83 (H5N2). Virology 1989;168:274-280.
20. Swayne DE, Pavade G, Hamilton K, Vallat B, Miyagishima K: Assessment of national strategies for control of high pathogenicity avian influenza and low pathogenicity notifiable avian influenza in poultry, with emphasis on vaccines and vaccination. OIE Rev Sci Tech 2011;30:839-870.
21. OIE: Highly pathogenic avian influenza, Mexico. OIE Disease Information [On-line] 2012.
22. Swayne DE: Impact of vaccines and vaccination on global control of avian influenza. Avian Dis 2012;56:in press.
23. Liu J, Chen P, Jiang Y, Wu L, Zeng X, Tian G, et al: A duck enteritis virus-vectored bivalent live vaccine provides fast and complete protection against H5N1 avian influenza virus infection in ducks. J Virol 2011;85:10989-10998.
24. Bublot M, Pritchard N, Swayne DE, Selleck P, Karaca K, Suarez DL, et al: Development and use of fowlpox vectored vaccines for avian influenza. Ann N Y Acad Sci 2006;1081:193-201.
25. Swayne DE: The role of vaccines and vaccination in high pathogenicity avian influenza control and eradication. Expert Rev Vaccines 2012;11:in press.
26. Stone H, Mitchell B, Brugh M: In ovo vaccination of chicken embryos with experimental Newcastle disease and avian influenza oil-emulsion vaccines. Avian Dis 1997;41:856-863.
27. Breedlove C, Minc JK, Tang DC, van Santen VL, van Ginkel FW, Toro H: Avian influenza adenovirus-vectored in ovo vaccination: target embryo tissues and combination with Marek's disease vaccine. Avian Dis 2011;55:667-673.
28. Schultz-Cherry S, Dybing JK, Davis NL, Williamson C, Suarez DL, Johnston R, Perdue ML: Influenza virus (A/HK/156/97) hemagglutinin expressed by an alphavirus replicon system protects chickens against lethal infection with Hong Kong-origin H5N1 viruses. Virology 2000;278:55-59.

29. Song H, Nieto GR, Perez DR: A new generation of modified live-attenuated avian influenza viruses using a two-strategy combination as potential vaccine candidates. J Virol 2007; 81:9238-9248.
30. Steel J, Burmakina SV, Thomas C, Spackman E, Garcia-Sastre A, Swayne DE, Palese P: A combination in-ovo vaccine for avian influenza virus and Newcastle disease virus. Vaccine 2008; 26:522-531.
31. Pavade G, Awada L, Hamilton K, Swayne DE: Analysis of economic indicators, poultry density and performance of veterinary services for control of high pathogenicity avian influenza in poultry. OIE Rev Sci Tech 2011;30:661-671.
32. Alexander DJ, Parsons G: Protection of chickens against challenge with virulent influenza A viruses of Hav5 subtype conferred by prior infection with influenza A viruses of Hsw1 subtype. Arch Virol 1980;66:265-269.
33. Hunt LA, Brown DW, Robinson HL, Naeve CW, Webster RG: Retrovirus-expressed hemagglutinin protects against lethal influenza virus infections. J Virol 1988;62:3014-3019.
34. Zhang W, Tu J, Zhao Z, Chen H, Jin M: The new temperature-sensitive mutation PA-F35S for developing recombinant avian live attenuated H5N1 influenza vaccine. Virol J 2012;9:97.
35. Park MS, Steel J, Garcia-Sastre A, Swayne D, Palese P: Engineered viral vaccine constructs with dual specificity: avian influenza and Newcastle disease. Proc Natl Acad Sci USA 2006;103:8203-8208.
36. Gao W, Soloff AC, Lu X, Montecalvo A, Nguyen DC, Matsuoka Y, et al: Protection of mice and poultry from lethal H5N1 avian influenza virus through adenovirus-based immunization. J Virol 2006;80:1959-1964.
37. Swayne DE, Suarez DL, Schultz-Cherry S, Tumpey TM, King DJ, Nakaya T, et al: Recombinant Paramyxovirus type 1-avian influenza-H7 virus as a vaccine for protection of chickens against influenza and Newcastle disease. Avian Dis 2003;47:1047-1050.
38. Ge J, Deng G, Wen Z, Tian G, Wang Y, Shi J, et al: Newcastle disease virus-based live attenuated vaccine completely protects chickens and mice from lethal challenge of homologous and heterologous H5N1 avian influenza viruses. J Virol 2007;81:150-158.
39. Rauw F, Palya V, Van BS, Welby S, Tatar-Kis T, Gardin Y, et al: Further evidence of antigenic drift and protective efficacy afforded by a recombinant HVT-H5 vaccine against challenge with two antigenically divergent Egyptian clade 2.2.1 HPAI H5N1 strains. Vaccine 2011;29:2590-2600.
40. Veits J, Luschow D, Kindermann K, Werner O, Teifke JP, Mettenleiter TC, Fuchs W: Deletion of the non-essential UL0 gene of infectious laryngotracheitis (ILT) virus leads to attenuation in chickens, and UL0 mutants expressing influenza virus haemagglutinin (H7) protect against ILT and fowl plague. J Gen Virol 2003;84:3343-3352.
41. Pavlova SP, Veits J, Mettenleiter TC, Fuchs W: Live vaccination with an H5-hemagglutinin-expressing infectious laryngotracheitis virus recombinant protects chickens against different highly pathogenic avian influenza viruses of the H5 subtype. Vaccine 2009;27:5085-5090.
42. Pan Z, Zhang X, Geng S, Cheng N, Sun L, Liu B, et al: Priming with a DNA vaccine delivered by attenuated Salmonella typhimurium and boosting with a killed vaccine confers protection of chickens against infection with the H9 subtype of avian influenza virus. Vaccine 2009;27:1018-1023.
43. Layton SL, Kapczynski DR, Higgins S, Higgins J, Wolfenden AD, Liljebjelke KA, et al: Vaccination of chickens with recombinant Salmonella expressing M2e and CD154 epitopes increases protection and decreases viral shedding after low pathogenic avian influenza challenge. Poult Sci 2009;88:2244-2252.
44. Chambers TM, Kawaoka Y, Webster RG: Protection of chickens from lethal influenza infection by vaccine-expressed hemagglutinin. Virology 1988;167:414-421.
45. Schultz-Cherry S, Dybing JK, Davis NL, Williamson C, Suarez DL, Johnston R, Perdue ML: Influenza virus (A/HK/156/97) hemagglutinin expressed by an alphavirus replicon system protects chickens against lethal infection with Hong Kong-origin H5N1 viruses. Virology 2000;278:55-59.
46. Crawford J, Wilkinson B, Vosnesensky A, Smith G, Garcia M, Stone H, Perdue ML: Baculovirus-derived hemagglutinin vaccines protect against lethal influenza infections by avian H5 and H7 subtypes. Vaccine 1999;17:2265-2274.
47. Kalthoff D, Giritch A, Geisler K, Bettmann U, Klimyuk V, Hehnen HR, et al: Immunization with plant-expressed hemagglutinin protects chickens from lethal highly pathogenic avian influenza virus H5N1 challenge infection. J Virol 2010;84:12002-12010.
48. Rao S, Kong WP, Wei CJ, Yang ZY, Nason M, Styles D, et al: Multivalent HA DNA haccination protects against highly pathogenic H5N1 avian influenza infection in chickens and mice. PLoS ONE 2008; 3(6):e2432.
49. Suarez DL, Schultz-Cherry S: The effect of eukaryotic expression vectors and adjuvants on DNA vaccines in chickens using an avian influenza model. Avian Dis 2000;44:861-868.

50. Swayne DE, Kapczynski DR: Vaccines, vaccination and immunology for avian influenza viruses in poultry, in Swayne DE (ed): Avian Influenza. Ames, Iowa, Blackwell Publishing,2008, pp 407-451.
51. Steensels M, Bublot M, Van Borm S, De Vriese J, Lambrecht B, Richard-Mazet A, et al: Prime-boost vaccination with a fowlpox vector and an inactivated avian influenza vaccine is highly immunogenic in Pekin ducks challenged with Asian H5N1 HPAI. Vaccine 2009;27:646-654.
52. Swayne DE: Principles for vaccine protection in chickens and domestic waterfowl against avian influenza: Emphasis on Asian H5N1 high pathogenicity avian influenza. Ann NY Acad Sci 2006;1081:174-181.
53. Maas R, Rosema S, van Zoelen D, Venema S: Maternal immunity against avian influenza H5N1 in chickens: limited protection and interference with vaccine efficacy. Avian Pathol 2011;40:87-92.
54. Swayne DE, Beck JR, Kinney N: Failure of a recombinant fowl poxvirus vaccine containing an avian influenza hemagglutinin gene to provide consistent protection against influenza in chickens preimmunized with a fowl pox vaccine. Avian Dis 2000;44:132-137.

David E. Swayne, DVN, PhD, dACVP, dACPV, Laboratory Director, Southeast Poultry Research Laboratory, USDA/Agricultural Research Service, 934 College Station Road, Athens, Georgia 30605 USA

Tel: 706-546-3433, fax: 706-546-3161

Email: David.Swayne@ars.usda.gov

Roth JA, Richt JA, Morozov IA (eds): Vaccines and Diagnostics for Transboundary Animal Diseases. Dev Biol (Basel). Basel, Karger, 2013, vol 135, pp 95-106.

Newcastle Disease: Progress and Gaps in the Development of Vaccines and Diagnostic Tools

C.L. Afonso[1], *P.J. Miller*[2]

[1] Southeast Poultry Research Laboratory (SEPRL), Agricultural Research Service (ARS), US Department of Agriculture (USDA), South Atlantic Area (SAA), Athens, GA, USA

[2] Veterinary Medical Officer, SEPRL, ARS, USDA, SAA, Athens, GA, USA

Key words: Newcastle disease, epidemiology, vaccines, diagnostics

Abstract: Newcastle disease (ND) is a contagious disease of birds that can have severe economic consequences for poultry producers, including a serious impact on the international trade of poultry and eggs. Newcastle disease virus (NDV) isolates are also called avian paramyxovirus serotype-1 isolates, but only infection with virulent NDV (vNDV) causes the disease. Virulent Newcastle disease virus (vNDV) isolates are distributed worldwide and have a high capacity to mutate, allowing the development of multiple vNDV genotypes evolving simultaneously at different locations. Large gaps in existing knowledge in the areas of epidemiology and evolution limit the possibilities to control the disease. Recurrent infection of poultry and wild birds allows the maintenance of a reservoir for the viruses; however, the role of wild birds and poultry in vNDV evolution is largely unknown. In the area of diagnostics, the performance of fast and accurate diagnostics methods is often affected by the evolution of viral genomes. Therefore, there is a need for the validation of multiple recently developed experimental tests and a need to develop additional fast and inexpensive diagnostic tests to be used in the field. In the area of vaccination, the development of inexpensive thermostable NDV vaccines and the development of vaccines capable of preventing viral replication are the highest priorities for endemic countries. In countries considered free of vNDV the development of low-cost vaccines that produce minimal vaccine reactions to prevent decreased productivity are higher priorities. Worldwide, better strategies that replace the culling of infected birds are needed to control outbreaks.

Introduction: the Disease and the Agent

Newcastle disease (ND) is a contagious and often fatal infectious viral disease affecting most species of birds. Because the disease can rapidly spread causing serious economic consequences to the poultry industry, as well as impacting the international trade of poultry and poultry products, ND is a disease reportable to the World Organization for Animal Health (OIE). Member countries of the OIE are required to report an outbreak within 24 hours after diagnosis. Death rates up to 100 percent in unvaccinated flocks are possible with ND outbreaks and vNDV is a major agro-bioterrorism threat agent for the United States (US) (http://www.fas.org/sgp/crs/terror/RL32521.pdf).

Virulent Newcastle disease viruses (vNDV) are defined as those isolates that produce a value equal to or greater than 0.7 in the intracerebral pathogenicity index assay (ICPI) in day-old chickens or those viruses that contain at least three basic amino acids in the fusion cleavage site positions 113-116 along with a phenylalanine in position 117 [1]. vNDV isolates are found worldwide in poultry and often in wild birds and are endemic in areas of Africa, Asia, the Middle East, and in some countries of North, Central and South America, causing recurrent outbreaks in poultry production facilities. vNDV isolates are ubiquitous in pigeons worldwide and are a constant threat due to their propensity to live near poultry facilities. Since the virus continues to evolve in wild and domestic birds, there is a constant threat that vNDV could be introduced into US poultry.

The virus is a member of the genus *Avulavirus* [2, 3] in the *Paramyxoviridae* family. It is a negative-sense, single stranded, non-segmented, enveloped RNA virus that is also known as avian paramyxovirus serotype-1 (APMV-1) [4]. The genome is composed of six genes encoding for their corresponding six proteins: nucleoprotein (NP), phosphoprotein (P), matrix (M), fusion (F), hemagglutinin-neuraminidase (HN), and the RNA polymerase (L) and the V protein produced as a translational frameshift [5]. The F protein cleavage site is the most critical site responsible for virulence; however, all genes are believed to be important for virus replication. The amino acid sequence at the F protein cleavage site of a vNDV is described as starting at position 113 as: 113,R-K/R-Q-K/R-R-F117 [1]. The less virulent viruses (loNDV) have fewer basic amino acids in those positions and a leucine instead of a phenylalanine at position 117. Using reverse genetic techniques it has been shown that the phenylalanine at position 117 and three basic amino acids surrounding position 114 are essential and sufficient for significant changes in virulence, i.e. localized versus systemic infection, causing either a low or a high pathogenic infection, respectively [6, 7].

Different genotypes of APMV-1 circulate worldwide. Although all NDV are members of APMV-1 and are of only one serotype, antigenic and genetic diversity is common [8, 9]. A stringent evaluation of phylogeny performed recently revealed the existence of a large number of distinct genotypes subdivided in two classes.

Class I viruses consist of only one genotype and class II viruses contain 15 genotypes [10]. Most poultry infections in the US are caused by lentogenic viruses of Class II genotype II, which include the low-virulence NDV vaccines used in the US such as the LaSota and B1 strains. Class II viruses also include the vaccines of genotype I used in Africa and Asia, as well as the neurotropic velogenic, virulent AMPV-chicken/US/(TX)GB/1948 strain isolated in the US, which is currently used to test vaccine efficacy in the US. The viruses of genotypes V, VI, and VII are virulent and are the most predominant, circulating worldwide. Viruses of these genotypes display the largest genetic diversity and can been divided in sub-genotypes.

Vaccination Practices

Vaccination combined with good biosecurity practices has been the recommended programme for control of ND for over 50 years [11]. Oil-emulsion vaccines are usually produced from inactivated NDV-infected allantoic fluid and applied via injection to individual birds. Live vaccines may be ingested or inhaled as part of mass application programmes in food, drinking water or by large droplet aerosol [5]. Although the current live vaccines protect against disease in specific pathogen free (SPF) chickens, there are still a significant number of outbreaks occurring in endemic countries suggesting deficiencies in either vaccines or vaccination methods.

Current NDV vaccines can be effective in preventing clinical diseases, and do considerably reduce viral shedding upon infection. However, commercial vaccines do not produce sterilizing immunity and cannot prevent viral replication; and vaccinated birds still shed vNDV at considerable levels. Antigenic differences between vaccines and currently circulating virulent strains may facilitate the escape and evolution of virulent NDV by allowing higher levels of viral shedding for heterologous virulent strains[12]. There is strong evidence that current vaccines prevent disease but cannot stop viral shedding [13-15]. Existence of field-isolates that escape neutralizing antibodies induced by vaccination has been reported [16]. NDV variants were isolated at multiple locations and time points in Korea with an increase in the frequency of isolation from 21% in 2002 to 85.7% in 2003-2006. These authors further identified one variant epitope in the HN gene and developed a recombinant vaccine utilizing the variant virus HN gene in a LaSota strain backbone [17]. In addition, evidence is emerging that using genotype-matched vaccines can significantly reduce viral shedding. More recently, antigenically matched vaccines have demonstrated increased capacity to prevent viral shedding of viruses of genotype VII and V respectively [14, 18].

The production of recombinant NDV isolates using reverse genetics has been reported by Peeters [19] and others [20-22]. Several NDV infectious clones are now available and are being used in pathogenicity studies. Our lab, and others, have created chimeric NDV isolates by replacing the HN gene or F and HN genes of loNDV or virulent NDV isolates with those from vNDV isolates and shown that virulence can also be modified by switching these two genes [14, 23-25].

Several other NDV recombinant vaccines have been developed. The surface glycoproteins, HN and F, are the most important antigens for induction of a protective immune response as demonstrated by protection elicited when the proteins F or HN were expressed either using a vaccinia virus vector [26, 27] or as a DNA vaccine [28, 29]. Vectored vaccines consisting of a suitable carrier virus (e.g. meleagrid herpesviruses) expressing one or more immunogenic NDV proteins (usually F and/or HN) have been developed [30, 31]. Protection has also been induced by delivery of combined HN and F in the form of a virosome vaccine created by removal of virus antigens other than HN and F [32]. Examples of vectored vaccines are recombinant viruses based on Vaccinia [26], Fowlpox virus [33], Pigeonpox virus [34], Meleagrid herpesvirus [30], Gallid herpesvirus-2 [29] and avian adeno-associated virus [35]. Most of these vaccines have not reached mass usage because of various reasons, including high cost for some vaccines, reduced protection, or issues associated with application. However, commercial producers in the US are increasingly beginning to use the Meleagrid herpesvirus (also known as HVT-vectored) NDV vaccines, because they do not produce the type of respiratory side effects that typically occurs in commercial poultry with traditional live NDV vaccines. Some other types of recombinant NDV-based vaccines are already commercially available in Mexico (http://www.engormix.com/iasa-investigacion-aplicada/genovax-sh10043_pr26921.htm); and others are being developed in Asia, mainly in China and Korea.

Diagnostic Methods

Molecular diagnostics is emerging as a viable alternative to classical diagnostics because of the ease of use, low cost and rapid results. Even though non-PCR based assays (e.g. virus isolation, or ICPI) are often not very sensitive and take a longer time, they are still considered to be gold standards. Therefore, a need remains for robust and standardized fast approaches to detect NDV. Most current rapid diagnostic methods are based on the use of reverse transcription followed by real time PCR (RT-PCR) and those have been reviewed recently by Hoffman and will not be further discussed here [36]. Of the many available molecular diagnostic methods, those based on RT-PCR seem to be the most accepted because of their speed and sensitivity [36].

Despite the availability of numerous molecular diagnostics for NDV, only one method of RT-PCR has undergone complete validation in the US and is widely used worldwide, with other methods used regionally or in specific occasions [37]. The Wise method, which is widely used in the US for surveillance and diagnostics, is based on the use of two tests: the matrix gene test, which is used to identify all APMV-1, and the fusion gene test, which is used to identify vNDV.

Gaps in Epidemiology and Evolution

Evidence suggests that the reservoir of virulent viruses across the planet is large; however, very little is known about NDV reservoirs. Based on the fact that over 250 avian species are known to be susceptible to infection with vNDV, the likelihood of wild birds and poultry maintaining vNDV as reservoirs is high. There is evidence of specific associations between genotypes and host species with virulent viruses; however, no evidence associating genotypes and host species has been found for the less studied and characterized loNDV isolates [38]. Repeated infection and recurrent outbreaks have been observed in cormorants with vNDV of subgenotype Va and pigeons are generally infected with viruses of genotype VI [38]. There is also now some evidence that NDV may have the capacity to expand the host range, as cormorant specific viruses have been recently isolated from other wild birds including gulls and pelicans [39]; however, it is not clear at the moment if those isolations represent isolated spillovers or the beginning of an expansion into new hosts.

An evolutionary assumption that can be made is that virulent NDV is maintained by replication in immunized poultry. This assumption is based not on compelling scientific evidence but merely on the lack of epidemiological evidence supporting the idea that virulent viruses are maintained in wild birds. Furthermore, poultry vNDV of newer genotypes have been shown to be genetically related to older viruses that caused earlier poultry outbreaks. Among vaccinated poultry, backyard poultry has the potential to be responsible for the continued existence of virulent viruses. Because of the wide distribution of backyard poultry in the developing world, and because the known deficiencies in vaccination for this production system, backyard poultry has the potential to constitute one of largest reservoir for virulent viruses. Since backyard poultry have varying levels of immunity, animals with low levels of antibodies may allow for more extensive virus replication and virus evolution. The existence of a larger number of outbreaks in countries with large population of backyard poultry appears to support this concept; however, no studies have been done to document the role of backyard poultry in the maintenance of NDV reservoirs. The frequent detection of virulent viruses of genotype VII in geese as well as the evidence that numerous species are susceptible to infection but do not show clinical signs, suggest that additional reservoirs are still to be found in wild birds.

There is also the potential risk that virulent viruses may arise by mutation from loNDV isolates. Mutation from loNDV to vNDV has been demonstrated to occur readily following intracerebral inoculation in chicken brains [6]; however, reversion to virulence is rare in nature. There have only been two known geographically and temporarily restricted outbreaks that were caused by virulent viruses that originated from a mutant loNDV that reverted to virulent. The best studied case is the outbreak caused by a change in the fusion cleavage site of a loNDV that circulated in birds in Australia [40], where the cleavage site changed from RRQKRL (West Sidney 2000 with ICPI of 0.41) to RRQGRF (1997) and then to RRQRRF (outbreak in 1998, ICPI 1.8). The second case is an outbreak that occured in Ireland in 1990 [41]. Furthermore, viruses of genotype II that resemble vaccine viruses but with an increased virulence have been isolated in China and Egypt; and it was recently found that the genome of NDV in turkeys within the US has mutated to become more like virulent viruses [42-44].

A previous study by Miller suggests that the annual rate of change of vNDV could be as high as ten times the rate of change of loNDV, suggesting that other selective pressures such as vaccination or increased host range may accelerate the rate of evolution of virulent viruses [45]. The prevalence of vNDV class II, genotypes (V-XV) in vaccinated poultry today [46], and the increased pathogenicity observed in recent isolates [47] suggest that the immune pressure from vaccination may be selecting variant forms of vNDV. This persistence of vNDV in poultry despite intensive vaccination efforts has been a recurrent phenomenon in endemic countries of Asia, Africa, Central and South America [46,48,49]. Some characteristics of poultry production contributing to virus evolution include: (1) genetic homogeneity of the host; (2) close animal-to-animal contact; and (3) vaccination programmes that provide selective immune pressures but do not prevent virus replication [50].

While evidence that genetic variants may have originated directly as a result of vaccination is missing, the phylogenetic analysis of the complete coding region of the fusion gene from recent Mexican vNDV suggests that those vNDV isolated between 1998 and 2006 formed a distinct group from viruses isolated previously [49]. Two groups of viruses separated into either an early (1998 - 2001) or a more recent group (2004 -2006). The ICPI and the mean death time (MDT) results (traditionally used to evaluate the pathogenicity of any new NDV isolate) suggested that some distinct selective pressures on the 1998–2001 viruses may have led to the appearance of the new, slightly more virulent group. The 1998–2001 isolates showed a 98% percent identity with other Mexican viruses reported between 1996 and 2001, but these were only 93–94% identical to the 2004-2006 isolates.

There is a need to increase the epidemiological database of circulating NDV in the US and around the world to avoid epidemiological gaps. Recently, two new genotypes have been reported in Peru and the Dominican Republican [51, 52]. However, the origin of these viruses remained elusive, thus confirming the large gap in epidemiological information for NDV. Most notable was the discovery of rare genotype XVI virus in the Dominican Republic. Birds on this island of the Caribbean may have carried viruses of this genotype unnoticed for the last 20 years. As new genotypes emerge worldwide, more questions will be raised on the ancestry of these viruses. The creation of large epidemiological databases will help with the identification of reservoirs, as well as the determination of risks of introducing virulent viruses into the US. It is also important to identify potential temporary carriers and avian species in which vNDV is asymptomatic. At the moment, only two confirmed wild bird reservoirs (pigeon and cormorant) have been documented, and these species seem to be reservoirs of just two genotypes [53, 54]. It is important to identify the avian species that are the reservoirs of all other genotypes.

Gaps in Diagnostics

There is a need to develop additional real time PCR tests for new genotypes of vNDV. Because of the rapid genomic changes of NDV, as well as the constant appearance of new variants, there is also a need to develop new procedures to improve the sensitivity and specificity of real-time RT-PCR. The failure of both the matrix United States Department of Agriculture (USDA) validated test and the fusion test to identify new vNDV attest to the need for new tests. Both assays have limitations caused by the constant genomic changes that occur during evolution of the viruses. The matrix test in general does not recognize almost 70% of viruses of class I [38]; and recently the matrix test has also been found to fail to recognize Class II virulent viruses from Pakistan [55]. The fusion test often fails to recognize viruses from pigeons (genotype VI) [54] and cormorants (Genotype Va) [39], and is a less reliable test because it is based on a more variable region of the genome. Recently vNDV isolates obtained from outbreaks of ND affecting cormorants and gulls in the states of Massachusetts, Maine, New Hampshire and Maryland were found to be closely related to the viruses that caused the ND outbreaks in Minnesota in 2008. In addition to the results obtained with the 2008 isolates, the standard USDA fusion (F) gene real-time reverse-transcription PCR (RRT-PCR) assay failed to detect the 2010 cormorant viruses. However, all viruses were detected by non-validated cormorant-specific F-gene RRT-PCR assay developed by the USDA as an alternate test [39]. Even when highly conserved regions are used, it is still possible to have failure of detection because of the genetic variation of NDV, highlighting the importance of constant surveillance and the re-assessment of current assays. However, manuscripts reporting new rapid diagnostic test have proliferated exponentially. Most new tests are based on molecular diagnostics and are either used only experimentally or not used at all in a diagnostic setting. From the practical point of view of controlling the disease in the field, there is also a need to develop diagnostic discovery research focused on flock side tests that can be used to make quick decisions in the field during an outbreak.

Conclusions: Gaps in Vaccination

The challenges in vaccination for developing countries differ significantly from the need in most industrialized nations. In the poorest countries of the developing world, the decreased ability to obtain vaccines and the absence of a cold chain to keep them viable are the most significant problems that prevent control of the disease [56, 57]. The absence of periodic vaccination causes recurrent waves of infection with virulent viruses that devastate backyard poultry production. One of the most pernicious effects of these outbreaks is the effect on the local entrepreneurial spirit, as such outbreaks discourage investment in the utilization of more efficient poultry equipment and installations. As a result, in the poorest countries most of the poultry production is confined to highly inefficient systems based on household production by women. Recently, the non-profit organization Galvmed (www.galvmed.org) has developed a new presentation of the Newcastle disease vaccine, in which smaller more-affordable amounts of fast-dissolving tablets are packaged into foil blister packs that can be stored for at least 24 hours at room temperature. The low cost and stability of this vaccine promises to dramatically improve access to vaccines for poor people by making them cheaper, easier to transport and more convenient to store and use. Each tablet, which is reconstituted in under 10 seconds in just 1 ml of water, will contain enough vaccine to protect 50 birds. Conventional vaccines are normally packed in glass vials, which are bulky, easily damaged, usually need to be stored under refrigeration and typically contain 500 or more doses.

In countries with middle incomes that can afford veterinary support for vaccination of large production facilities (e.g. Mexico, Indonesia and Egypt), deficiencies in the control of the disease are variable. Most of those deficiencies are related to the existence of endemic viruses in small-scale production facilities and backyard poultry units, or to failures of vaccination caused by multiple factors. As a result, viruses continue to evolve and antigenic differences between vaccine and circulating viruses are increasing. There are an increasing number of reports suggesting that classical vaccines do not perform well to prevent viral replication; however, there is no solid evidence suggesting failure of classical vaccines to protect against morbidity and mortality caused by new vNDV. In these countries, increased surveillance and better management practices are likely to improve productivity significantly; moreover, the use of homologous vaccines as tools to reduce viral replication and shedding is already being implemented in Mexico and appears to have had some beneficial effects (Dr. Eduardo Lucio Decanini, IASA, personal communication).

In the developed world and in countries in which virulent NDV only occasionally infects poultry, vaccines needs are different; and research is being conducted with the following objectives in mind: 1) Prevention of transmission; 2) Cross-protection against variant strains; 3) Maternal antibody override; 4) Induction of protection with a single dose; 5) Low cost mass vaccination; 6) Increased safety and minimal side effects; and 7) Differentiation of infected from vaccinated animals (DIVA). A different set of needs for the developed world has led to the development of vectored vaccines that satisfy some but not all of these requirements. Although current live vaccines protect against disease, they may not be as effective in controlling environmental viral loads against emerging Asian and South American strains; and there is clear evidence that these current vectored vaccines cannot completely prevent viral replication and shedding [14]. The development of new vectors with increased antigen delivery capacity and minimal side effects needs to be investigated. Another area of research in vaccination that is likely to have a significant impact on the efficiency and cost of vaccination is the development of improved inactivated vaccine formulations (new emulsion candidates) that increase immunogenicity (single dose) but also reduce the withdrawal time following application.

The two most pursued areas in NDV vaccine research are: 1) Development of improved immunogenic responses based on the development of vaccines with better antigenic matching; and 2) Development of vaccines that cause reduced post vaccination side effects. The first objective is being pursued using reverse genetics to develop live attenuated viruses that are identical to circulating viruses (except for the fusion protein cleavage site) or by developing viruses that contain the LaSota vaccine backbone and the F and HN genes from circulating viruses. The second objective is being pursued through the use of subunit vaccines based on delivery of antigens using alternative vectors (HVT, Fowlpox, Adenoviruses). Some of this vaccine research is already having considerable economic impact in countries in which the risk of challenge with virulent NDV is low and with the added advantage of allowing differentiation of infected from vaccinated animals (DIVA). Since subunit vaccines do not induce as strong immune responses as live vaccines do, they are often used in combination with live vaccines.

Vaccination practices in the developed world, in middle and low income countries are all improving, adapting to specific situations and local challenges. This improvement is being made possible by a remarkably diverse programme of scientific research. Substantive additional funding is required; and the sooner such funding is made available, the sooner innovative solutions will be implemented, reducing the existing gaps in vaccination research and application. Such additional investment in facing ND would lead to an immediate reduction in the extent of the disease, resulting in the improved health of poultry in many nations and the expansion of disease-free international trade in poultry and poultry products.

Acknowledgements

SEPRL research was funded by USDA ARS CRIS number 6612-32000-064. We would like to thank the Southeash Poultry Research Laboratory (SEPRL), animal support personnel, SEPRL sequencing facilities, and Tim Olivier & Dawn Williams-Coplin for technical support on the research leading to this manuscript.

References

1 OIE: Manual of diagnostic tests and vaccines for terrestrial animals: mammals, birds and bees, in Biological Standards Commission. Paris, World Organization for Animal Health, 2012, vol 1, part 2, Chapter 2.03.14., pp.1-19.

2 Fauquet CM, Fargette D: International Committee on Taxonomy of Viruses and the 3,142 unassigned species. Virol J 2005;2:64.

3 Mayo MA: A summary of taxonomic changes recently approved by ICTV. Arch Virol 2002;147(8):1655-1656.

4 Lamb R, Parks GD: Paramyxoviridae: The viruses and their replication, In Fields B, Knipe, DM and Howley PM (eds): Fields virology. Philadelphia, Lippincott Williams and Wilkins,2007, vol 1, pp 1449-1496.

5 Alexander DJ, Senne DA: Newcastle disease, other avian paramyxoviruses, and pneumovirus infections, in Saif YM, Fadly AM, Glisson JR, McDougald LR, Nolan LK, Swayne DE (eds); Diseases of Poultry. 12th edidtion, Ames, Iowa State University Press, 2008, pp 75-116.

6 De Leeuw OS, Hartog L, Koch G, Peeters BPH: Effect of fusion protein cleavage site mutations on virulence of Newcastle disease virus: Non-virulent cleavage site mutants revert to virulence after one passage in chicken brain. J Gen Virol 2003;84:475-484.

7 Glickman RL, Syddall RJ, Iorio RM, Sheehan JP, Bratt MA: Quantitative basic residue requirements in the cleavage-activation site of the fusion glycoprotein as a determinant of virulence for Newcastle disease virus. J Virol 1988;62(1):354-356.

8 Aldous EW, Mynn JK, Banks J, Alexander DJ: A molecular epidemiological study of avian paramyxovirus type 1 (Newcastle disease virus) isolates by phylogenetic analysis of a partial nucleotide sequence of the fusion protein gene. Avian Pathol 2003;32(3):239-257.

9 Alexander DJ, Swayne DE: Newcastle disease virus and other avian paramyxoviruses, in Swayne DE, Glisson JR, Jackwood MW, Pearson JE, Reed WM (eds): A Laboratory Manual for the Isolation and Identification of Avian Pathogens. Kennett Square, PA, The American Association of Avian Pathologists, 1998,4th ed, pp 156-163.

10 Diel DG, da Silva LHA, Liu H, Wang Z, Miller PJ, Afonso C: Genetic diversity of avian paramyxovirus type 1: Proposal for a unified nomenclature and classification system of Newcastle disease virus genotypes. Infect Genet Evol 2013; in press.

11 Senne DA, King DJ, Kapczynski DR: Control of Newcastle disease by vaccination. Dev Biol 2004;119:165-170.

12 Miller PJ, King DJ, Afonso CL, Suarez DL: Antigenic differences among Newcastle disease virus strains of different genotypes used in vaccine formulation affect viral shedding after a virulent challenge. Vaccine 2007;25(41):7238-7246.

13 Kapczynski DR, King DJ: Protection of chickens against overt clinical disease and determination of viral shedding following vaccination with commercially available Newcastle disease virus vaccines upon challenge with highly virulent virus from the California 2002 exotic Newcastle disease outbreak. Vaccine 2005;23:3424-3433.

14 Miller PJ, Estevez C, Yu Q, Suarez DL, King DJ: Comparison of viral shedding following vaccination with inactivated and live Newcastle disease vaccines formulated with wild-type and recombinant viruses. Avian Dis 2009;53(1):39-49.

15 Utterback WW, Schwartz JH: Epizootiology of velogenic viscerotropic Newcastle disease in Southern California, 1971-1973. J Am Vet Med Assoc 1973;163:1080-1088.

16 Cho SH, Kwon HJ, Kim TE, Kim JH, Yoo HS, Kim SJ: Variation of a Newcastle disease virus hemagglutinin-neuraminidase linear epitope. JCM 2008;46(4):1541-1544.

17 Cho SH, Kwon HJ, Kim TE, Kim JH, Yoo HS, Park MH, et al: Characterization of a recombinant Newcastle disease vaccine strain. Clin Vaccine Immunol 2008;15(10):1572-1579.

18 Hu S, Ma H, Wu Y, Liu W, Wang X, Liu Y, Liu X: A vaccine candidate of attenuated genotype VII Newcastle disease virus generated by reverse genetics. Vaccine 2009; 27(6):904-910.

19 Peeters BP, De Leeuw OS, Koch G, Gielkens AL: Rescue of Newcastle disease virus from cloned cDNA: Evidence that cleavability of the fusion protein is a major determinant for virulence. J Virol 1999;73(6):5001-5009.

20 Krishnamurthy S, Huang ZH, Samal SK: Recovery of a virulent strain of Newcastle disease virus from cloned cDNA: Expression of a foreign gene results in growth retardation and attenuation. Virology 2000;278(1):168-182.

21 Nakaya T, Cros J, Park MS, Nakaya Y, Zheng HY, Sagrera A, et al: Recombinant Newcastle disease virus as a vaccine vector. J Virol 2001;75(23):11868-11873.

22 Romer-Oberdorfer A, Mundt E, Mebatsion T, Buchholz UJ, Mettenleiter TC: Generation of recombinant lentogenic Newcastle disease virus from cDNA. J Gen Virol 1999;80:2987-2995.

23 De Leeuw OS, Koch G, Hartog L, Ravenshorst N, Peeters BPH: Virulence of Newcastle disease virus is determined by the cleavage site of the fusion protein and by both the stem region and globular head of the haemagglutinin-neuraminidase protein. J Gen Virol 2005;86(Pt. 6):1759-1769.

24 Huang ZH, Panda A, Elankumaran S, Govindarajan D, Rockemann DD, Samal SK: The Hemagglutinin-Neuraminidase Protein of Newcastle disease virus determines tropism and virulence. J Virol 2004;78(8):4176-4184.

25 Susta L, Miller PJ, Afonso CL, Estevez C, Yu Q, Zhang J, Brown CC: Pathogenicity evaluation of different Newcastle disease virus chimeras in 4-week-old chickens. Trop Anim Health Prod 2010;42(8):1785-1795.

26 Meulemans G, Letellier C, Gonze M, Carlier MC, Burny A: Newcastle disease virus F glycoprotein expressed from a recombinant vaccinia virus vector protects chickens against live-virus challenge. Avian Pathol 1988;17(4) 821-828.

27 Nishino Y, Niikura M, Suwa T, Onuma M, Gotoh B, Nagai Y, Mikami T: Analysis of the protective effect of the haemagglutinin neuraminidase protein in Newcastle disease virus infection. J Gen Virol 1991;72:1187-1190.

28 Heckert RA, Elankumaran S, Oshop GL, Vakharia VN: A novel transcutaneous plasmid-dimethylsulfoxide delivery technique for avian nucleic acid immunization. Vet Immunol Immunopathol 2002;89(1-2):67-81.

29 Sakaguchi M, Nakamura H, Sonoda K, Okamura H, Yokogawa K, Matsuo K, Hira K: Protection of chickens with or without maternal antibodies against both Marek's and Newcastle diseases by one-time vaccination with recombinant vaccine of Marek's disease virus type 1. Vaccine 1998;16(5):472-479.

30 Heckert RA, Riva J, Cook S, McMillen J, Schwartz RD: Onset of protective immunity in chicks after vaccination with a recombinant herpesvirus of turkeys vaccine expressing Newcastle disease virus fusion and hemagglutinin-neuraminidase antigens. Avian Dis 1996;40(4):770-777.

31 Morgan RW, Gelb J, Schreurs CS, Lutticken D, Rosenberger JK, Sondermeijer PJA: Protection of chickens from Newcastle and Marek's diseases with a recombinant herpesvirus of turkeys vaccine expressing the Newcastle disease virus fusion protein. Avian Dis 1992;36(4):858-870.

32 Kapczynski DR, Tumpey TM: Development of a virosome vaccine for Newcastle disease virus. Avian Dis 2003;47(3):578-587.

33 Boursnell MEG, Green PF, Samson ACR, Campbell JIA, Deuter A, Peters RW, Millar NS, Emmerson PT, Binns MM: A recombinant fowlpox virus expressing the hemagglutinin-neuraminidase gene of Newcastle disease virus (NDV) protects chickens against challenge by NDV. Virology 1990;178:297-300.

34 Letellier C, Burny A, Meulemans G: Construction of a pigeonpox virus recombinant: Expression of the Newcastle disease virus (NDV) fusion glycoprotein and protection of chickens against NDV challenge. Arch Virol 1991;118(1-2):43-56.

35 Perozo F, Villegas P, Estevez C, Alvarado IR, Purvis LB, Saume E: Avian adeno-associated virus-based expression of Newcastle disease virus hemagglutinin-neuraminidase protein for poultry vaccination. Avian Dis 2008;52(2):253-259.

36 Hoffmann B, Beer M, Reid SM, Mertens P, Oura CA, van Rijn PA, et al: A review of RT-PCR technologies used in veterinary virology and disease control: Sensitive and specific diagnosis of five livestock diseases notifiable to the World Organisation for Animal Health. Vet Microbiol 2009;139(1-2):1-23.

37 Wise MG, Suarez DL, Seal BS, Pedersen JC, Senne DA, King DJ, et al: Development of a real-time reverse-transcription PCR for detection of Newcastle disease virus RNA in clinical samples. J Clin Microbiol 2004;42(1):329-338.

38 Kim LM, King DJ, Curry PE, Suarez DL, Swayne D, Stallknecht DE, et al: Phylogenetic diversity among low-virulence Newcastle disease viruses from waterfowl and shorebirds and comparison of genotype distributions to those of poultry-origin isolates. J Virol 2007; 81(22):12641-12653.

39 Diel DG, Miller PJ, Wolf PC, Mickley RM, Musante AR, Emanueli DC, et al: Characterization of Newcastle disease viruses isolated from cormorant and gull species in the United States in 2010. Avian Dis 2012;56(1):128-133.

40 Gould AR, Kattenbelt JA, Selleck P, Hansson E, Della-Porta A, Westbury HA: Virulent Newcastle disease in Australia: Molecular epidemiological analysis of viruses isolated prior to and during the outbreaks of 1998-2000. Virus Res 2001; 77(1):51-60.

41 Alexander DJ, Campbell G, Manvell RJ, Collins MS, Parsons G, McNulty MS: Characterisation of an antigenically unusual virus responsible for two outbreaks of Newcastle disease in the Republic of Ireland in 1990. Vet Rec 1992;130(4):65-68.

42 Tan L-T, Xu H-Y, Wang Y-L, Qin Z-M, Sun L, Liu W-J, Cui Z-Z: Molecular characterization of three new virulent Newcastle disease virus variants isolated from China. J Clin Microbiol 2008;46(2):750-753.

43 Mohamed MH, Kumar S, Paldurai A, Samal SK: Sequence analysis of fusion protein gene of Newcastle disease virus isolated from outbreaks in Egypt during 2006. Virol J 2011;8:237.

44 Miller PJ, Killian ML, Pederson JC, Afonso CL: Avian paramyxovirus serotype 1 strains of low virulence with unusual fusion protein cleavage sites isolated from poultry species [abstract], in Pathologists AAA (ed): American Veterinary Medical Association. St. Louis MO AAAP, 2011.

45 Miller PJ, Kim LM, Ip HS, Afonso CL: Evolutionary dynamics of Newcastle disease virus. Virology 2009;391(1):64-72.

46 Czegledi A, Ujvari D, Somogyi E, Wehmann E, Werner O, Lomniczi B: Third genome size category of avian paramyxovirus serotype 1 (Newcastle disease virus) and evolutionary implications. Virus Res 2006;120(1-2):36-48.

47 Susta L, Miller PJ, Afonso CL, Brown CC: Clinicopathological characterization in poultry of three strains of Newcastle disease virus isolated from recent outbreaks. Vet Pathol 2011;48(2):349-360.

48 Alexander DJ: Newcastle Disease. Brit Poultry Sci 2001;42(1):5-22.

49 Perozo F, Merino R, Afonso CL, Villegas P, Calderon N: Biological and phylogenetic characterization of virulent Newcastle disease virus circulating in Mexico. Avian Dis 2008;52(3):472-479.

50 Higgins DA, Shortridge KF: Newcastle disease in tropical and developing countries, in Alexander, D J (ed): Developments in veterinary virology: Newcastle disease XIII, Boston MA, Kluwer, 1988, pp 273-302 .

51 Diel DG, Susta L, Cardenas Garcia S, Killian ML, Brown CC, Miller PJ, Afonso CL: Complete genome and clinicopathological characterization of a virulent Newcastle disease virus isolated from poultry in South America. J Clin Microbiol 2011;50(2):378-387.

52 Courtney SC, Gomez D, Susta L, Hines N, Pedersen JC, Miller PJ, Afonso CL: Complete genome sequencing of a novel Newcastle disease virus isolate circulating in layer chickens in the Dominican Republic. J Virol 2012;86(17):9550.

53 Kim LM, King DJ, Guzman H, Tesh RB, Travassos da Rosa AP, Bueno R, Jr, et al: Biological and phylogenetic characterization of pigeon paramyxovirus serotype-1 circulating in wild North American pigeons and doves. J Clin Microbiol 2008;46(10):3303-3310.

54 Rue CA, Susta L, Brown CC, Pasick JM, Swafford SR, Wolf PC, et al: Evolutionary changes affecting rapid diagnostic of 2008 Newcastle disease viruses isolated from double-crested cormorants. J Clin Microbiol 2010;48(7):2440-2448.

55 Khan TA, Rue CA, Rehmani SF, Ahmed A, Wasilenko JL, Miller PJ, Afonso CL: Phylogenetic and biological characterization of Newcastle disease virus isolates from Pakistan. J Clin Microbiol 2010;48(5):1892-1894.

56 Henning J, Morton J, Pym R, Hla T, Meers J: Evaluation of strategies to improve village chicken production-controlled field trials to assess effects of Newcastle disease vaccination and altered chick rearing in Myanmar. Prev Vet Med 2009;90(1-2):17-30.

57 Spradbrow P: Appropriate vaccination and therapies for rural poultry flocks in developing countries and their relevance to developed countries. World Poultry Sci J 2005;61(1):47-54.

Claudio L. Afonso, Southeast Poultry Research Laboratory, Agricultural Research Service, US Department of Agriculture, South Atlantic Area, 934 College Station Road, Athens, GA 30605.

Tel: + 1-706-546-3642/3645/3434, fax: + 1-706-546-3161

Email: claudio.afonso@ars.usda.gov

Roth JA, Richt JA, Morozov IA (eds): Vaccines and Diagnostics for Transboundary Animal Diseases. Dev Biol (Basel). Basel, Karger, 2013, vol 135, pp 107-116.

Novel Approaches to Foot-and-Mouth Disease Vaccine Development

A. Ludi[1,2], L. Rodriguez[1]

1 United States Department of Agriculture, Agricultural Research Service, Foreign Animal Disease Research Unit, Plum Island Animal Disease Center, Orient Point, New York, NY, USA

2 Oak Ridge Institute for Science and Education, Plum Island Animal Disease Center Research Participation Program, Oak Ridge, TN, USA

Key words: vaccine development, foot-and-mouth disease, novel approaches

Abstract: The need for better foot-and-mouth disease (FMD) vaccines is not new, a report from the Research Commission on FMD, authored by F. Loeffler and P. Frosch in 1897, highlighted the need for developing a vaccine against FMD and qualified this as a devastating disease causing "severe economic damage to the country's agriculture" [1]. Inactivated antigen vaccines have been available since the early 1900s and have been instrumental in eradicating FMD from parts of the world and repressing clinical disease in others. However, these vaccines require using live virulent FMDV for manufacturing, fail to prevent infection resulting in the establishment of carrier animals, require multiple vaccination schedules (every six months) and have limited coverage to the specific serotype and in many cases subtype of FMDV. Therefore, FMD vaccinology continues to be a very active research field. Research-based novel vaccine approaches over the last decade have resulted in at least one novel molecular vaccine being licensed for emergency use in the US and multiple other vaccine approaches being actively pursued as alternatives to current vaccines. Here we will review and update the main research efforts on FMD vaccines, including subunit and peptide vaccines, DNA vaccines, empty capsid vaccines (directly delivered or vector delivered), novel inactivated antigen production platforms and live attenuated vaccines. Each of these approaches will be discussed in terms of their safety and efficacy characteristics, product transition feasibility as well as their applicability to global control and eradication efforts.

INTRODUCTION

"We have a large selection of safe and effective vaccines to control the next FMD outbreak in mice and guinea pigs" L. Rodriguez, 2012.

Foot-and-mouth disease (FMD) remains the main sanitary barrier to the commerce of animals and animal products worldwide. It is one of the main threats to the livestock economies of FMD-free countries where recent disease incursions have had serious economic impact [2, 3]. In addition, FMD continues to have a negative effect on the livestock of endemic regions, limiting animal production in some of the poorest regions of the world with high food insecurity. Some of these regions are also undergoing population expansions that demand animal protein sources and FMD is an important limiting factor to livestock production and contributes to food insecurity in these regions [4]. Vaccination remains the main control method for FMD. Inactivated antigen vaccines are effective in preventing clinical disease but do not always prevent viral infection and replication at the primary sites resulting in persistent infections in approximately 50% of vaccinated cattle exposed to FMD virus (FMDV). Another problem with currently available vaccines is their narrow coverage, limited to the serotype and in many cases the subtype of FMDV. This causes problems when new strains emerge or are introduced which can potentially result in serious vaccine coverage failures. A recent example of this problem is the introduction of serotype SAT 2 into Northern Africa where vaccine formulations did not include that serotype. Additionally, one of the most important shortcomings of the current FMD vaccines is the short duration of immunity, which is limited to 4-6 months in animals after a single vaccination [5].

There is a clear need for FMD vaccines with improved profiles, capable of preventing infection. These vaccines should not only suppress clinical disease, but also have broader antigenic coverage at least within the most problematic serotypes and have the necessary antigenic markers to allow easy differentiation between infected and vaccinated animals (DIVA). Recent publications have reviewed state of the art FMD vaccine research and development [5-7]; here we will update previous reviews and focus on the research aspects for the development of the next generation of FMD vaccines. We will restrict our review to those vaccines that have been tested for efficacy in livestock as previous experience has shown that it is difficult to translate efficacy data from laboratory animals (mice, guinea pigs) to cattle and swine.

INACTIVATED ANTIGEN VACCINES

FMD vaccines currently consist of inactivated purified antigen (killed virus depleted of non-structural proteins) and various formulations of adjuvants. After manufacturing, antigen concentrates are usually kept frozen in liquid nitrogen. FMD vaccines are usually formulated from these frozen antigen stocks. The World

Organisation for Animal Health Manual suggests a six protective dose 50% (PD_{50}) per cattle delivered most commonly by intramuscular injection. However, three PD_{50} is the industry standard for routine vaccines and these provide protection from challenge within four to seven days post-vaccination in cattle, swine and sheep [8-11]. Protective immune responses require booster vaccination every six months for at least the first two years of age with subsequent annual revaccination [12-14].

It was recently shown that alternate routes of delivery such as intra-dermal deposition through needle-free devices provide full protection in cattle at 1/16 of the standard antigen load [15].

Various adjuvants are currently used, with the double oil-emulsion being the most common. Double oil-emulsion vaccines are thought to provide a stronger and longer antibody response than water-in-oil single emulsion vaccines [16, 17]. The aluminum hydroxide-saponin formulated vaccines have the advantage of being easier to produce; however, they can cause granulomas at the inoculation site and are not effective in swine [18-21]. These aluminum hydroxide-saponin formulated vaccines are mostly used in ruminants.

Inactivated vaccines are currently the only type of vaccines used for control of FMD. They are useful in controlling the disease through decrease in clinical disease and virus amplification. However, vaccination does not prevent infection, resulting in persistent infection in some vaccinated cattle exposed to FMDV. Additionally, vaccine production requires the growth of large volumes of fully virulent viruses, and escape from laboratories with manufacturing facilities has occurred recently [22].

Novel Inactivated Antigen Production Platform

There is a clear need in industry for safer FMD vaccine production. A reasonable approach to safer vaccines is to use attenuated FMD viruses that are capable of growth in-vitro but are harmless to animals. It has been well documented that FMD viruses lacking the leader protease (Lpro) are attenuated in cattle and swine [23-25]. Taking advantage of this knowledge, a recent report utilized cDNA technology to derive a FMD virus lacking the leader protease and the epitopes in non-structural proteins, 3B and 3D, recognized by specific diagnostic monoclonal antibodies [26]. This report describes an inactivated vaccine candidate utilizing this technology, a leaderless A24 FMD virus containing negative markers in proteins 3B and 3D (FMD-A24-LL3B3D). Epitopes deleted in 3B and 3D were recognized by particular monoclonal antibodies. Using these deleted epitopes and corresponding monoclonal antibodies, companion diagnostic ELISA tests were developed that allow serological differentiation between infected and vaccinated animals (DIVA). The genetic stability of the mutations introduced in the Lpro, 3B and 3D was confirmed by 15 serial passages in BHK-21 cells with no detectable changes. The safety of FMD-A24-LL3B3D was tested both in cattle and swine. Cattle inoculated with this virus had no clinical signs and no viral shedding or transmission to animals in direct contact. These animals did not show any signs of fever, viremia, clinical manifestation or shedding of virus in saliva or air samples. These animals also did not develop significant levels of neutralizing antibodies during the course of the experiment. Binary ethylenimine (BEI) inactivated FMD-A24-LL3B3D vaccine proved to be as effective as commercially available FMD vaccines in cattle when challenged with the parental virus. Similarly swine vaccinated with leaderless A-24 virus and challenged showed no clinical signs, no fever and no transmission to other swine. One of the benefits of this vaccine platform is that it allows for easy exchange of the capsid region to any FMDV serotype and subtype desired, due to unique restriction endonuclease sites flanking the capsid-coding region in the infectious cDNA plasmid. Also animals vaccinated with this marker virus platform can be easily distinguished from naturally infected animals by competitive ELISA tests developed even after multiple vaccinations have taken place. Having marker vaccines also simplifies vaccine production by not having to remove non-structural proteins in order to maintain the DIVA capability.

DNA Vaccines

DNA vaccines are thought to be a great alternative to the current vaccine as they do not require high containment facilities for production, have a relatively stable shelf-life, allow for rapid incorporation of emerging field strain sequences, can incorporate marker genes, and can co-express multiple antigenic sites [6]. However, DNA vaccines are limited in that they require multiple doses, the addition of adjuvant and cytokines, and have a decreased and short lived antibody response. Fowler et al. were able, for the first time, to show partial protection using a DNA vaccine in cattle [27]. This had previously only been demonstrated in sheep and pigs [6]. The DNA vaccine produced contained the capsid region P1 and the non-structural proteins 2A, 3C and 3D. This vaccine was administered using electroporation delivery device and included the immune modulator bovine GM-CSF. A booster was given 21 days after the first vaccine and the animals had a protein boost of 146S FMD virus antigen and non-structural 3D protein 42 days after primary vaccination. The animals were then challenged 133 days after the primary vaccine. When this DNA vaccine was administered without electroporation, only 25% (1 out of 4) animals were protected compared to 75% (3 out of 4 cattle) with electroporation. The addition of GM-CSF reduced clinical disease and virus secretion from nasal swabs.

A second recent study in swine used a DNA vaccine to incorporate one B-cell FMDV epitope, two T-cell FMDV epitopes and the capsid region of FMDV [28]. Two different vaccines were made with one containing a strong signal peptide. Neither of the DNA vaccines was able to prevent disease. In fact the group of swine that received the strong signal peptide vaccine had more severe disease than that seen for the positive control animals. The reason for this is unknown, but the authors suggest cytoxic T lymphocytes (CTL) could be responsible. The same laboratory also tested a DNA vaccine (pCMV-APCH1BTT), which contained an FMDV B and T-cell epitope fused to a single-chain variable fragment of the 1F12 mouse monoclonal antibody that recognized Class-II Swine Leukocyte antigen [29]. Using this DNA vaccine two of four swine were protected against challenge while the other half was partially protected. Swine were vaccinated every two weeks for a total of three doses and were challenged fifteen days post-vaccination. Although none of the DNA vaccines described to date are practical for routine use, they represent an effort to show protection with this method in large animals and perhaps will allow for future studies toward more practical DNA vaccines.

Peptide Vaccines

Peptide vaccines have several advantages over the traditional inactivated vaccines. Peptide vaccines do not involve infectious virus, can easily be stored because they consist of denatured protein and the purity of the target protein is greater than 95% [30]. Several peptide vaccines have been tested over the years showing various degrees of efficacy. Early studies incorporated what is thought to be the most antigenic site, the G-H loop of VP1, into peptide vaccines but these studies showed inconsistent results in cattle [31]. By incorporating both a T-helper site and consensus residues in the hypervariable positions, higher protection was achieved in swine following challenge [32]. However, these results could not be repeated in cattle; no protection from clinical disease was observed after challenge of vaccinated animals [33]. More recently, a dendrimeric peptide containing one copy of an FMDV T-cell epitope branching into four copies of a B-cell epitope has shown improvement over the linear peptide [34]. Swine vaccinated with this dendrimeric peptide developed comparable immune responses as those infected with wild type FMDV.

A recent study uses three repeats of residues 141 to 160 and 200 to 213 of VP1 from the O/China/99 strain to form a peptide vaccine [30]. The peptide was coupled with a swine immunoglobulin G heavy-chain constant region which was thought to prolong the half-life of the antigenic epitope in vivo and improve the ability to deliver peptide to major histocompatibility complex in situ. These multiple epitope recombinant vaccines conferred complete protection against challenge with 10^3 50% swine infective doses. (This is tenfold less than the OIE recommended dose for challenging pigs, which is 10^4 $TCID_{50}$ titrated on swine cells). The anti-FMDV specific antibody titers were not significantly different from the traditional vaccine 30 days post-vaccination. Additionally the PD_{50} was 6.47, which is higher than what is recommended by the World Organisation for Animal Health. Despite these reports of protective responses in farm animals, the use of peptide vaccines is of concern since it has been shown previously that the FMD virus can easily circumvent immune responses induced by peptides.

In Vitro Expressed Virus-like Particles (VLP)

FMDV virus-like particles (VLPs) have the characteristics and antigenicity of the parental virus but are unable to replicate as they are devoid of RNA. Baculovirus and other similar vectors encoding the genetic information for these VLPs have the ability to express and process the viral capsid polyprotein into cleaved products that then assemble into VLPs in a variety of infected cells. Several expression systems for the production of these VLPs have been investigated. These include mammalian cells lines, baculovirus/insect cells and larvae system, various species of yeast and bacteria.

Little work on these VLPs has been carried out in livestock and what has been done shows inconsistent results. Baculovirus-derived as well as *E. coli*-derived experimental VLP vaccines have shown some protection against clinical disease in swine but failed to protect against viral replication [35]. Another report has shown that by using baculovirus and silkworm larvae, VLP can assemble in larvae lysates [36]. The diluted supernatant from lysing the silk hemolymph was used as a vaccine and induced protection in cattle that were challenged 21 days post-vaccination. A recent study in cattle by the same authors showed good protection against the parental virus (6.5 PD_{50}) and field virus (5.2 PD_{50}) on challenge with 10^3 BID_{50} given 21 dpi in cattle [37]. These results are promising and deserve further study to determine if baculovirus infected insect larvae are a feasible vaccine production platform for commercial use.

Viral Vector Vaccines Delivering VLP

Unlike the delivery of vaccines containing protein synthesized outside the host, viral vectors are used to deliver genetic information to express and process all of FMDV structural proteins with the aim of synthesizing virus-like particles inside the vector-infected cells. Several viral vectors have been reported to deliver FMD capsids including herpesvirus [38], poxvirus [39] and the human defective adenovirus 5 [40] vectors.

The hAd5-FMD vaccine has been the most successful viral vector vaccine and has been shown to have similar efficacy in cattle as inactivated vaccines. The vaccine has shown to be protective 7 days post-vaccination in both cattle and swine [40-42]. The advantage of this vaccine is that it does not require high biological containment facilities for manufacturing since the virus is defective and does not replicate except in cells expressing specific Ad5 viral proteins. It also does not require adaptation of field strains to vaccine production cells, since the capsid protein of the desired vaccine strain can be readily cloned into the Ad5 vector. The hAd5-FMD vaccine has recently been modified [43]. With the addition of the full length non-structural protein 2B, cattle developed significantly higher levels of both CD4+ and CD8+ T-cells responding to FMDV antigen stimulation [44]. The addition of 2B has also been shown to improve efficacy of a serotype O1Campos vaccine which is of significance because this serotype O requires a four to five times higher antigenic payload than serotype A to induce protective immune responses [43]. The Ad5-A24 vaccine is currently being manufactured in experimental batches and tested in the United States where it has recently obtained a conditional license for emergency use.

A recent study has used rabbit haemorrhagic disease virus (RHDV) as a vaccine vector to deliver FMDV VLPs to swine [45]. This vector contained the T-cell epitope of the 3A protein of FMDV. Although the swine were not challenged they were able to show the activation of immature porcine bone marrow-derived dendritic cells in vitro. Peripheral blood mononuclear cells (PBMC) were collected two and four weeks after immunization to analyze IFN-γ using ELISPOT. It was found that vaccinated pigs generated FMDV-specific IFN-γ producing cells against the 3A and the vector RHDV-VLP. IgA and IgG levels were also induced against the RHDV-VLP.

Modified Live Vaccines

There has not been significant research reported recently on live-attenuated virus (LAV) FMD vaccines likely because previous attempts have shown limited success due to unstable phenotype, differences in attenuation among various livestock species, inconsistency in inducing protective immune response, and concerns of reversion to virulence. However, based on functional genomics, several regions of the FMDV genome have been identified that are determinants of viral virulence. Specifically the leader protease (Lpro) has been shown to be critical for viral virulence and when removed, renders the virus non-virulent to cattle. However, these leaderless viruses are so highly attenuated that they do not replicate enough in the host to induce a protective immune response [23].

Recent work has revisited the potential of leader mutated FMDV as candidate modified live vaccines. Specifically, a recent manuscript described that by inserting two mutations in the Lpro coding region in a motif known as SAF-A/B, Acinus and PIAS (SAP), FMDV was completely attenuated in swine [46]. Swine vaccinated with A12-SAP (10^7 PFU/animal) were completely protected (i.e. no detectable clinical signs, viremia or virus shedding) from challenge with A12 wild type virus as early as two days post-inoculation and for at least twenty-one days post-inoculation [47]. The success of this vaccine is likely due to the ability of the virus to achieve limited replication in vaccinated animals because the two mutations in the SAP region impair the ability of Lpro to inhibit the host innate response. Animals inoculated with A12-SAP did develop significant levels of FMDV-specific neutralizing antibodies starting 7 dpi (peak at 14 dpi) without clinical signs or viremia. Lastly, tissue culture passage of the SAP-mutant virus displays stability for at least twelve passages. The SAP mutant is one potential live attenuated vaccine (LAV) candidate that looks promising. However, further research is necessary to explore attenuating factors in the FMDV genome that could be utilized for the development of stable and efficient attenuated vaccines.

Research Gaps and Future Work

There are major research gaps that remain in order to develop better FMD vaccines. There is a need for vaccines that are easy and safe to produce, easy to deliver and that confer both short and long term immunity to a broad antigenic spectrum of viral strains within each serotype. The capacity to safely manufacture vaccine is critical particularly for FMDV-free countries and countries that are in the process of FMDV eradication. There is a clear need for FMD vaccines that induce long-lasting immunity, as opposed to current vaccines that confer only 4-6 months protection. Although current vaccines effectively prevent clinical disease, they fail to prevent infection and as a result up to 50% of vaccinated protected cattle that are subsequently exposed to FMDV become chronic carriers. We need to understand the early events of FMDV pathogenesis and develop vaccines that can prevent infection of the primary infection site(s) hence preventing persistent infection. Finally there is a need for vaccines that are safe and inexpensive to produce, are thermostable and can be easily delivered to endemic regions. These new-generation vaccines are certainly within reach and are needed in order to achieve the goal of global control and eventual eradication of FMDV.

References

1 Brown F: The history of research in foot-and-mouth disease. Virus Res 2003;91(1):3-7.

2 Foot-and-mouth disease taking toll in Koreas. J Am Vet Med Assoc 2011;Sect. 966.

3 Thompson D, Muriel P, Russell D, Osborne P, Bromley A, Rowland M, et al: Economic costs of the foot and mouth disease outbreak in the United Kingdom in 2001. Rev Sci Tech 2002;21(3):675-687.

4 Vallat B, Lubroth J: The Global Foot and Mouth Disease Control Strategy - strengethening animal health systems through imporved control of major diseases. 2012; Available from: http://www.fao.org/docrep/015/an390e/an390e.pdf.

5 Rodriguez LL, Gay CG: Development of vaccines toward the global control and eradication of foot-and-mouth disease. Expert Rev Vaccines 2011;10(3):377-387.

6 Fowler VL, Barnett PV: Progress in the development of DNA vaccines against foot-and-mouth disease. Expert Rev Vaccines 2012;11(4):481-493.

7 Rodriguez LL, Grubman MJ: Foot and mouth disease virus vaccines. Vaccine 2009;27 Suppl 4:D90-94.

8 Salt JS, Barnett PV, Dani P, Williams L: Emergency vaccination of pigs against foot-and-mouth disease: protection against disease and reduction in contact transmission. Vaccine 1998;16(7):746-754.

9 Golde WT, Pacheco JM, Duque H, Doel T, Penfold B, Ferman GS, et al: Vaccination against foot-and-mouth disease virus confers complete clinical protection in 7 days and partial protection in 4 days: Use in emergency outbreak response. Vaccine 2005;23(50):5775-5782.

10 Brehm KE, Kumar N, Thulke HH, Haas B: High potency vaccines induce protection against heterologous challenge with foot-and-mouth disease virus. Vaccine 2008;26(13):1681-1687.

11 Barnett PV, Carabin H: A review of emergency foot-and-mouth disease (FMD) vaccines. Vaccine 2002;20(11-12):1505-1514.

12 Parida S: Vaccination against foot-and-mouth disease virus: strategies and effectiveness. Expert Rev Vaccines 2009;8(3):347-365.

13 Mattion N, Goris N, Willems T, Robiolo B, Maradei E, Beascoechea CP, et al: Some guidelines for determining foot-and-mouth disease vaccine strain matching by serology. Vaccine 2009;27(5):741-747.

14 Cox SJ, Aggarwal N, Statham RJ, Barnett PV: Longevity of antibody and cytokine responses following vaccination with high potency emergency FMD vaccines. Vaccine 2003;21(13-14):1336-1347.

15 Pandya M, Pacheco JM, Bishop E, Kenney M, Milward F, Doel T, et al: An alternate delivery system improves vaccine performance against foot-and-mouth disease virus (FMDV). Vaccine 2012;30(20):3106-3111.

16 Hunter P: The performance of southern African territories serotypes of foot and mouth antigen in oil-adjuvant vaccines. Rev Sci Tech 1996;15(3):913-922.

17 Iyer AV, Ghosh S, Singh SN, Deshmukh RA: Evaluation of three 'ready to formulate' oil adjuvants for foot-and-mouth disease vaccine production. Vaccine 2000;19(9-10):1097-1105.

18 Valtulini S, Macchi C, Ballanti P, Cherel Y, Laval A, Theaker JM, et al: Aluminium hydroxide-induced granulomas in pigs. Vaccine 2005;23(30):3999-4004.

19 Michelsen E, Maerki E: [Laboratory and field methods for the demonstration of anioactive detergents in surface, ground and sewage water]. Mitt Geb Lebensmittelunters Hyg 1961;52:557-571.

20 McKercher PD, Callis JJ: Foot-and-mouth disease: a vaccine study. Dev Biol Stand 1976;35:123-133.

21 Doel TR: FMD vaccines. Virus Res 2003;91(1):81-99.

22 Cottam EM, Wadsworth J, Shaw AE, Rowlands RJ, Goatley L, Maan S, et al: Transmission pathways of foot-and-mouth disease virus in the United Kingdom in 2007. PLoS Pathog 2008;4(4):e1000050.

23 Brown CC, Piccone ME, Mason PW, McKenna TS, Grubman MJ: Pathogenesis of wild-type and leaderless foot-and-mouth disease virus in cattle. J Virol 1996;70(8):5638-5641.

24 Mason PW, Piccone ME, McKenna TS, Chinsangaram J, Grubman MJ: Evaluation of a live-attenuated foot-and-mouth disease virus as a vaccine candidate. Virology 1997;227(1):96-102.

25 Chinsangaram J, Mason PW, Grubman MJ: Protection of swine by live and inactivated vaccines prepared from a leader proteinase-deficient serotype A12 foot-and-mouth disease virus. Vaccine 1998;16(16):1516-1522.

26. Uddowla S, Hollister J, Pacheco JM, Rodriguez LL, Rieder E: A safe foot-and-mouth disease vaccine platform with two negative markers for differentiating infected from vaccinated animals. J Virol 2012;86(21)11675-11685.

27 Fowler V, Robinson L, Bankowski B, Cox S, Parida S, Lawlor C, et al: A DNA vaccination regime including protein boost and electroporation protects cattle against foot-and-mouth disease. Antiviral Res 2012;94(1):25-34.

28 Ganges L, Borrego B, Fernandez-Pacheco P, Revilla C, Fernandez-Borges N, Dominguez J, et al: DNA immunization of pigs with foot-and-mouth disease virus minigenes: From partial protection to disease exacerbation. Virus Res 2011;157(1):121-125.

29 Borrego B, Argilaguet JM, Perez-Martin E, Dominguez J, Perez-Filgueira M, Escribano JM, et al: A DNA vaccine encoding foot-and-mouth disease virus B and T-cell epitopes targeted to class II swine leukocyte antigens protects pigs against viral challenge. Antiviral Res 2011;92(2):359-363.

30 Shao JJ, Wong CK, Lin T, Lee SK, Cong GZ, Sin FW, et al: Promising multiple-epitope recombinant vaccine against foot-and-mouth disease virus type O in swine. Clin Vaccine Immunol 2011;18(1):143-149.

31 Taboga O, Tami C, Carrillo E, Nunez JI, Rodriguez A, Saiz JC, et al: A large-scale evaluation of peptide vaccines against foot-and-mouth disease: lack of solid protection in cattle and isolation of escape mutants. J Virol 1997;71(4):2606-2614.

32 Wang CY, Chang TY, Walfield AM, Ye J, Shen M, Chen SP, et al: Effective synthetic peptide vaccine for foot-and-mouth disease in swine. Vaccine 2002;20(19-20):2603-2610.

33 Rodriguez LL, Barrera J, Kramer E, Lubroth J, Brown F, Golde WT: A synthetic peptide containing the consensus sequence of the G-H loop region of foot-and-mouth disease virus type-O VP1 and a promiscuous T-helper epitope induces peptide-specific antibodies but fails to protect cattle against viral challenge. Vaccine 2003;21(25-26):3751-3756.

34 Cubillos C, de la Torre BG, Jakab A, Clementi G, Borras E, Barcena J, et al: Enhanced mucosal immunoglobulin A response and solid protection against foot-and-mouth disease virus challenge induced by a novel dendrimeric peptide. J Virol 2008;82(14):7223-7230.

35 Grubman MJ, Lewis SA, Morgan DO: Protection of swine against foot-and-mouth disease with viral capsid proteins expressed in heterologous systems. Vaccine 1993;11(8):825-829.

36 Li Z, Yi Y, Yin X, Zhang Z, Liu J: Expression of foot-and-mouth disease virus capsid proteins in silkworm-baculovirus expression system and its utilization as a subunit vaccine. PLoS One 2008;3(5):e2273.

37 Li ZY, Yin XP, Yi YZ, Li XR, Li BY, Lan X, et al: FMD subunit vaccine produced using a silkworm-baculovirus expression system: Protective efficacy against two type Asia1 isolates in cattle. Vet Microbiol 2011;149(1-2):99-103.

38 Hong Q, Qian P, Li XM, Yu XL, Chen HC: A recombinant pseudorabies virus co-expressing capsid proteins precursor P1-2A of FMDV and VP2 protein of porcine parvovirus: a trivalent vaccine candidate. Biotechnol Lett 2007;29(11):1677-1683.

39 Zheng M, Jin N, Zhang H, Jin M, Lu H, Ma M, et al: Construction and immunogenicity of a recombinant fowlpox virus containing the capsid and 3C protease coding regions of foot-and-mouth disease virus. J Virol Methods 2006;136(1-2):230-237.

40 Moraes MP, Mayr GA, Mason PW, Grubman MJ: Early protection against homologous challenge after a single dose of replication-defective human adenovirus type 5 expressing capsid proteins of foot-and-mouth disease virus (FMDV) strain A24. Vaccine 2002;20(11-12):1631-1639.

41 Pacheco JM, Brum MC, Moraes MP, Golde WT, Grubman MJ: Rapid protection of cattle from direct challenge with foot-and-mouth disease virus (FMDV) by a single inoculation with an adenovirus-vectored FMDV subunit vaccine. Virology 2005;337(2):205-209.

42 Grubman MJ, Moraes MP, Schutta C, Barrera J, Neilan J, Ettyreddy D, et al: Adenovirus serotype 5-vectored foot-and-mouth disease subunit vaccines: the first decade. Future Virology 2010;5(1):51-64.

43 Moraes MP, Segundo FD, Dias CC, Pena L, Grubman MJ: Increased efficacy of an adenovirus-vectored foot-and-mouth disease capsid subunit vaccine expressing nonstructural protein 2B is associated with a specific T cell response. Vaccine 2011;29(51):9431-9440.

44 Pena L, Moraes MP, Koster M, Burrage T, Pacheco JM, Segundo FD, et al: Delivery of a foot-and-mouth disease virus empty capsid subunit antigen with nonstructural protein 2B improves protection of swine. Vaccine 2008;26(45):5689-5699.

45 Crisci E, Fraile L, Moreno N, Blanco E, Cabezon R, Costa C, et al: Chimeric calicivirus-like particles elicit specific immune responses in pigs. Vaccine 2012;30(14):2427-2439.

46 de los Santos T, Segundo FD, Zhu J, Koster M, Dias CC, Grubman MJ: A conserved domain in the leader proteinase of foot-and-mouth disease virus is required for proper subcellular localization and function. J Virol 2009;83(4):1800-1810.

47 Segundo FD, Weiss M, Perez-Martin E, Dias CC, Grubman MJ, Santos Tde L: Inoculation of swine with foot-and-mouth disease SAP-mutant virus induces early protection against disease. J Virol 2011;86(3):1316-1327.

Luis Rodriguez, United States Department of Agriculture, Agricultural Research Service, Foreign Animal Disease Research Unit, Plum Island Animal Disease Center, Orient Point, New York, NY, USA
Tel.: +1 631 323 3364, fax: +1 631 323 3006
E-mail: luis.rodriguez@ars.usda.gov

Roth JA, Richt JA, Morozov IA (eds): Vaccines and Diagnostics for Transboundary Animal Diseases. Dev Biol (Basel). Basel, Karger, 2013, vol 135, pp 117-123.

Diagnosis of Foot-and-Mouth Disease

D.J. Paton, D.P. King

The Pirbright Institute, Pirbright, Woking, Surrey, UK

Key words: Foot-and-mouth disease, diagnosis, penside tests, contingency planning

Abstract: Foot-and-mouth disease virus (FMDV) exists as multiple serotypes and strains that infect a range of cloven-hoofed animals with variable severity. Clinical diagnosis reinforced by diagnostic tests support timely intervention, whilst virus characterisation helps trace routes of spread and select appropriate vaccine strains. To speed up and simplify diagnosis, penside tests have recently been developed. Serology is used to identify undisclosed infection and substantiate freedom from infection and specific tests are needed to detect infected animals in vaccinated populations. Serology is also used to estimate post-vaccinal population immunity. Contingency plans are required to enable countries to scale up diagnosis at short notice. Improvements are needed in preclinical and penside diagnosis and in our ability to model vaccine effectiveness.

INTRODUCTION

Foot-and-mouth disease (FMD) is caused by an RNA virus, FMDV, in the family *Picornaviridae* and affects domestic and wild animals including cattle, pigs, sheep, goats, Asian and African buffalo, Bactrian camels, wild boar, deer and antelopes [1]. There are seven FMDV serotypes and multiple and constantly evolving sub-serotype strains, with particular serotypes and strains occurring in different parts of the world. Infection gives rise to a range of clinical outcomes with cattle and pigs tending to be worst affected, especially high productivity animals kept intensively. The disease is characterised by epithelial vesicles that rupture and erode, particularly affecting the mouth, tongue, feet and udder. Chronic lameness, heat sensitivity and milk loss may follow. Young animals may develop acute myocarditis and die suddenly. FMD can spread quickly between animals and farms by direct animal contacts and indirectly via fomites, aerosols and animal products. Infection of wild animals is usually self limiting, but African buffalo provide a long-term reservoir for FMDV. Developed countries have mostly eradicated the disease and maintain freedom by strict controls on imports of animals and their products from infected countries, making FMD status a constraint to international trade. Due to high costs and lack of adequate veterinary infrastructure, developing countries have difficulty implementing animal import and movement controls or affording mass vaccination for their susceptible livestock populations. Consequently, they tend to suffer from cyclical epidemics associated with the spread of newly emerging serotypes and strains of FMDV.

The lesions associated with full-blown cases of FMD are fairly characteristic, but can be mistaken for conditions caused by other vesicular viruses, e.g. swine vesicular disease and vesicular stomatitis viruses. Additionally, animals may show little or no obvious signs due to low susceptibility or partial immunity. Consequently, a range of diagnostic tests are needed to confirm clinical cases and to identify recovered animals or those that are not showing characteristic clinical signs of infection. FMD-free countries need contingency plans to rapidly detect incursions of FMD, as delay in diagnosis increases the likelihood of large-scale outbreaks. Rapid confirmation is essential in support of movement controls and culling strategies aimed at breaking transmission. Therefore, arrangements are needed to transport samples to laboratories in the shortest possible time or else to conduct testing locally. Confirmation of the first case, especially in a previously FMD-free country (or zone) has the potential to detrimentally impact the trading status, and will therefore need more extensive work-up compared to the requirements for subsequent secondary cases. Where FMD is endemic, diagnosis may be complicated by the presence of multiple serotypes of circulating virus. Vaccination also complicates diagnosis as tests are then required to differentiate between vaccinal immunity and that caused by infection.

Infection and Immune Dynamics

Virus excretion starts shortly before the onset of clinical signs and persists for several days thereafter [2]. Vesicular fluid and epithelium are rich sources of FMDV and are the specimens of choice for agent detection. Nasal and oropharyngeal secretions, breath, blood and milk also contain virus in the acute stages of the infection. FMDV may persist for prolonged periods in the pharynx of ruminants but not pigs (sampled with a probang cup), but the level of virus declines over time and may be intermittently detectable. Antibodies to the structural proteins (SP) of FMDV appear rapidly and can be detected from ~3 days after first appearance of lesions. Antibodies to non-structural viral proteins (NSP) appear two-three days later. Vaccination elicits antibodies mainly or entirely against SP, depending upon the extent to which NSP have been removed during vaccine purification.

Confirmation of Suspected Cases

Thorough clinical examination is important and the tongue and all feet should be carefully inspected. An estimate should be made of the age of any lesions found and a photograph may be taken and referred to an experienced diagnostician [3]. Thermal imaging can detect heat associated with acute foot lesions and could help prioritise suspicious cases for initial examination [4].

The presence of FMDV can be confirmed by tests for viral proteins and RNA as well as by isolation of cytopathic virus in cell cultures. Antigen detection ELISA [5] can detect and serotype the virus present in vesicular samples and from other matrices after amplification in cell cultures. Lateral flow devices (LFD) have been developed to simplify this process and to facilitate penside testing; panserotype-specific LFDs and type-specific LFDs for some serotypes have been recently made available [6, 7]. These and other devices can also be used to stabilise viral RNA for submission and downstream amplification/sequencing in a specialist laboratory. Amplification of viral RNA or of live virus are more sensitive techniques that can be applied to all sample types, including probang and aerosol samples that may contain only trace amounts of virus. Serotyping can be achieved with molecular assays by use of serotype-specific primers and sequencing the amplified viral RNA gives additional strain information [8]. Both RT-PCR [9] and isothermal tests, such as RT-LAMP [10], can provide highly sensitive assays for FMD and with suitably labelled primers the test products can be detected by LFD. Recently, we have shown that both aerosol samples and epithelial suspensions can be tested with high sensitivity in a RT-LAMP/LFD combination assay without prior RNA extraction, further simplifying the process for penside diagnosis.

Due to the rapidity with which animals seroconvert to FMDV, serology is often useful in estimating the extent and duration of infection, even in a herd or flock where clinically affected animals are the main target for confirmatory diagnosis. Use of a combination of tests, for IgM and for SP and NSP specific antibodies, can aid this process [11].

Virus Characterisation

Phylogenetic studies can identify the strain of virus involved in an outbreak and help trace where it may have come from. Serological tests for SP antibodies are used to determine which available vaccine is most closely related and to estimate the degree of match as a predictor of vaccine efficacy. However, such tests are difficult to standardise and interpret and other factors such as vaccine potency will also impact on efficacy. In vivo vaccination and challenge tests can be performed, but may be too slow to inform the decision on emergency vaccine selection and use. The revolution in nucleic acid sequencing has facilitated the sequencing of the entire genome of FMDV isolates on a routine basis and the improvement in resolution between closely related viruses has enabled the order of farm-farm spread to be deduced from phylogeny [12]. Further precision and inferences may be achieved with next generation sequencing. Sequencing of the FMDV capsid encoding genes is under development as a means of predicting the antigenic match between vaccines and field isolates [13].

Detection of Cases in the Absence of Clinical Signs

Preclinical diagnosis could enhance the efficacy of controls, such as culling or future use of antiviral treatments [14]. Low levels of virus can be detected in blood and some mucosal surfaces prior to the onset of disease. Using RT-PCR testing of cattle blood samples collected every other day, a proactive approach was employed in the UK in 2007 to screen premises at high risk due to proximity to earlier outbreaks [9]. However, a less laborious sampling and testing procedure would be highly desirable for use in large scale outbreaks. Other possible non-invasive methods to screen large numbers of animals at an early stage of infection include bulk milk testing and targeted air sampling.

Clinical inspection and serology remain the principal tools for assessing the extent of undisclosed infection and substantiating absence of infection. Effort should be focussed on populations at highest risk. Different serological tests are available with attributes favouring particular circumstances. NSP tests can detect infection by all serotypes and are relatively specific for infection rather than vaccination. The problem of excluding vaccinal antibodies is much less where highly purified vaccines are used in an emergency, in previously unvaccinated animals, as non-specific reactions are mainly a feature of low purity vaccines and multiple vaccination [15]. Serotype-specific IgA tests can also be used to detect evidence of local FMDV replication in the oropharynx as an indicator of infection and not vaccination [16].

The possibility that recovered domestic animals may initiate new outbreaks as a result of long-term virus carriage and shedding is controversial and if it occurs at all is a very rare event. Recovered animals can be detected by serology but confirmation of virus carriage requires probang sampling and testing, and negative findings require repeat sampling and testing to exclude carriage with confidence. Paired serology within the associated herd or flock can be used to find out if virus is circulating. Recently developed IgA tests are relatively specific for ongoing virus replication and therefore have potential to differentiate carrier and non-carrier convalescent animals.

Field Measures of Vaccine Efficacy

The effectiveness of FMD vaccination programmes can be monitored by estimating vaccine coverage and population immunity. The latter is done by serological surveys for SP antibodies. However, there is a lack of consistency in the approaches used, with variation in survey design (inclusion of only vaccinated animals or not; differing intervals between vaccination and sampling) test selection (use of a test virus homologous to the vaccine strain or to the circulating field strain against which protection is sought) and interpretation (cut-off value for acceptance of immune protection in individual animals and at the level of the population). Observational studies of FMD vaccine effectiveness, to directly compare the protection afforded in the field to vaccinated versus unvaccinated animals have been rarely conducted.

Testing Capability and Capacity

Maintaining a capability (including physical resources, manpower and accredited systems) to deal with the diagnostic burdens that accompany a large epidemic of FMD in a normally disease-free country is a challenge. A cadre of suitably trained reserve staff should be established along with contingency plans to recruit additional testing laboratories, for example to undertake testing for serosurveillance. Diagnostic tests reliant on amplification of large quantities of live FMDV should only be undertaken in laboratories with suitable biocontainment facilities and procedures. However, risk assessment may identify tests that can be performed safely in laboratories with less stringent biocontainment using procedures that do not require live virus to be handled and that mitigate the risk from the samples themselves. Countries lacking high containment facilities should have arrangements to refer samples to laboratories able to conduct a full work up when required, but some on-the-spot capability is desirable [17].

Conclusions

A wide range of tests are available to support different aspects of FMD surveillance and control. Guidance on test selection and interpretation is provided in the OIE Manual of Diagnostic Techniques for Terrestrial Animals [18] whilst the framework in which testing is conducted in support of outbreak controls is described in the EC Directive [19].

Better methods for preclinical FMD diagnosis would be beneficial in support of stamping out policies for FMD control. Penside tests are becoming available but require further refinement and commercialisation, for example, a simple low-cost platform to heat samples for isothermal methods (such as LAMP) would be helpful. Issues related to logistics, statutory control and quality control for penside testing are under debate [20]. The importance of detecting carrier animals remains controversial as well as technically difficult and countries that go to great lengths to do so may be penalised in relation to trade. Vaccine selection remains a difficult area, and even if it is possible to identify which vaccine strain would be most appropriate, it is not yet feasible to model the implications of a certain level of match and potency in relation to disease control efficacy. In endemically infected countries with complex epidemiological situations where sequential infection occurs with multiple serotypes and vaccines of uncertain purity are used, interpretation of serology to monitor infection or vaccinal immunity remains challenging and longitudinal studies could help elucidate what is possible.

Acknowledgements

Thanks are due for discussions over many years with colleagues at Pirbright and elsewhere. Funding for diagnostic developments at The Pirbright Institute has been provided by BBSRC and multiple grants from Defra (currently SE1127, SE4211) and the European Commission (currently FMD Disconvac, Rapidia-Field). DP is a Jenner Investigator.

References

1 Thomson GR, Bastos ADS: Foot and mouth disease: In Coetzer JAW & Tustin RC (eds): Infectious diseases of livestock, Cape Town, Oxford University Press, 2004, Vol 2, pp 1324-1365.

2 Alexandersen S, Zhang Z, Donaldson AI, Garland AJ: The pathogenesis and diagnosis of foot-and-mouth disease. J Comp Pathol 2003;129(1):1-36.

3 Anonymous: Foot-and-mouth disease. Ageing of lesions. 1986, revised 2005. http://archive.defra.gov.uk/foodfarm/farmanimal/diseases/atoz/fmd/documents/ageing-lesions.pdf

4 Gloster J, Ebert K, Gubbins S, Bashiruddin J, Paton DJ: Normal variation in thermal radiated temperature in cattle: implications for foot-and-mouth disease detection. BMC Vet Res 2011;7:73.

5 Ferris NP, Dawson M: Routine application of enzyme-linked immunosorbent assay in comparison with complement fixation for the diagnosis of foot-and-mouth and swine vesicular diseases. Vet Microbiol 1988;16(3):201-209.

6 Ferris NP, Nordengrahn A, Hutchings GH, Reid SM, King DP, Ebert K, et al: Development and laboratory validation of a lateral flow device for the detection of foot-and-mouth disease virus in clinical samples. J Virol Methods 2009;155(1):10-17.

7 Jiang T, Liang Z, Ren WW, Chen J, Zhi XY, Qi GY, et al: A simple and rapid colloidal gold-based immunochromatographic strip test for detection of FMDV serotype A. Virol Sin 2011;26(1):30-39.

8 Knowles NJ, Samuel AR: Molecular epidemiology of foot-and-mouth disease virus. Virus Res 2003;91:65-80.

9 Reid S.M, Ebert K, Bachenek-Bankowska K, Batten C, Sanders A, Wright C, et al: Performance of real-time RT-PCR for the detection of foot-and-mouth disease virus during field outbreaks in the United Kingdom in 2007. J Vet Diagn Invest 2009;21(3):321-330.

10 Dukes JP, King DP, Alexandersen S: Novel reverse transcription loop-mediated isothermal amplification for rapid detection of foot-and-mouth disease virus. Arch Virol 2006;151:1093-1106.

11 Ryan E, Gloster J, Reid S, Li Y, Ferris NF, Waters R, et al: Clinical and laboratory investigations of the foot-and-mouth disease epidemic in the United Kingdom in 2007. Vet Rec 2008;63:139-147.

12 Cottam EM, Wadsworth J, Shaw AE, Rowlands RJ, Goatley L, Maan S, et al: Transmission Pathways of Foot-and-Mouth Disease Virus in the United Kingdom in 2007. PLoS Pathogens 2008;4:1-8.

13 Reeve R, Blignaut B, Esterhuysen JJ, Opperman P, Matthews L, Fry EE, et al: Sequence-based prediction for vaccine strain selection and identification of antigenic variability in foot-and-mouth disease virus. PLoS Comput Biol 2010;6(12):e1001027.

14 Charleston B, Bankowski BM, Gubbins S, Chase-Topping ME, Schley D, Howey R, et al: Relationship between clinical signs and transmission of an infectious disease and the implications for control. Science 2011;332(6030):726-729.

15 Uttenthal A, Parida S, Rasmussen TB, Paton DJ, Haas B, Dundon WG: Strategies for differentiating infection in vaccinated animals (DIVA) for foot-and-mouth disease, classical swine fever and avian influenza. Expert Rev Vaccines 2010;9(1):73-87.

16 Parida S, Anderson J, Cox SJ, Barnett PV, Paton DJ: Secretory IgA as an indicator of oro-pharyngeal foot-and-mouth disease virus replication and as a tool for post vaccination surveillance. Vaccine 2006;24:1107-1116.

17 Paton DJ, Ferris NP, Hutchings GH, Li Y, Swabey K, Keel P, et al: Investigations into the cause of foot-and-mouth disease virus seropositive small ruminants in Cyprus during 2007. Transbound Emerg Dis 2009;56(8):321-328.

18 Anonymous: OIE Manual of Standards for Diagnostic Tests and Vaccines. World Organisation for Animal Health, Paris, France. Volume 1, Chapter 2.1.5. Foot and mouth disease, 2012. http://www.oie.int/fileadmin/Home/eng/Health_standards/tahm/2.01.05_FMD.pdf

19 Anonymous: Council Directive 2003/85/EC on Community measures for the control of foot-and-mouth disease repealing Directive 85/511/EEC and Decisions 89/531/EEC and 96/665/EEC and amending Directive 92/46/EEC. Article 21 & 45. Off J Eur Union L306, 2003.

20 Sammin D, Ryan E, Ferris N, King D, Zientara S, Haas B, Yadin H, et al: Options for decentralised testing of suspected secondary outbreaks of foot-and-mouth disease. Transbound Emerg Dis 2010;57(4):237-243.

Prof David J Paton, The Pirbright Institute, Pirbright, Woking, Surrey, GU24 ONF, UK.
Email: david.paton@pirbright.ac.uk

Roth JA, Richt JA, Morozov IA (eds): Vaccines and Diagnostics for Transboundary Animal Diseases. Dev Biol (Basel). Basel, Karger, 2013, vol 135, pp 125-138.

Passive Immunization and Active Vaccination against Hendra and Nipah Viruses

C.C. Broder

Department of Microbiology and Immunology, Uniformed Services University, Bethesda, MD, USA

Key words: Hendra, Nipah, vaccine, glycoprotein, monoclonal antibody

Abstract: Hendra virus and Nipah virus are viral zoonoses first recognized in the mid and late 1990's and are now categorized as the type species of the genus *Henipavirus* within the family *Paramyxoviridae*. Their broad species tropism together with their capacity to cause severe and often fatal disease in both humans and animals make Hendra and Nipah "overlap agents" and significant biosecurity threats. The development of effective vaccination strategies to prevent or treat henipavirus infection and disease has been an important area of research. Here, henipavirus active and passive vaccination strategies that have been examined in animal challenge models of Hendra and Nipah virus disease are summarized and discussed.

Introduction

Hendra virus (HeV) and Nipah virus (NiV) are zoonoses and recent additions to the family *Paramyxoviridae* comprising the genus *Henipavirus*. They are membrane-enveloped, single-stranded negative-sense RNA viruses whose natural reservoir hosts are known to be several species of fruit bats (flying foxes) of the genus *Pteropus*. In bats the henipaviruses do not cause apparent disease, but HeV and NiV are distinguished from all other known paramyxoviruses in their uniquely broad tropism and can also cause an often fatal infection in multiple other mammalian hosts including horses, pigs, cats, dogs, guinea pigs, hamsters, ferrets, monkeys and humans, which in addition to bats, spans 6 mammalian Orders (reviewed in [1]). HeV appeared first in eastern Australia in 1994 and was transmitted to humans from infected horses, while NiV emerged a few years later in 1998-99 in peninsular Malaysia and was primarily transmitted to humans from infected pigs, but several other animal species were also found infected (reviewed in [2]). Accordingly, the henipaviruses can be amplified and cause disease in animals followed by subsequent transmission to humans, where infection presents as a severe respiratory illness and/or encephalitis often with high case fatality rates (reviewed in [3]).

Over the past fifteen years a remarkable quantity of information has been uncovered on the molecular and cellular biology of the henipaviruses, their prevalence in various bat populations and geographic distribution, the epidemiology and risk factors associated with transmission, the diagnosis of infection and as well as the pathological manifestations observed following infection in a wide variety of hosts, and in the development of varied and effective countermeasures. Several of these areas of henipavirus research have been recently and extensively reviewed elsewhere [4-7]. Nevertheless, their broad species tropism together with their capacity to cause severe and often fatal disease in both humans and animals classify HeV and NiV as "overlap agents" and significant biosecurity threats resulting from natural outbreaks, laboratory accidents, or deliberate dispersal. Here, the current status in the development of effective vaccination strategies, which have been tested in animal models, to prevent or treat henipavirus infection and disease in people and economically important livestock is summarized. Relevant key literature cited is limited and not intended to be exhaustive.

Hendra Virus and Nipah Virus Spillovers

HeV was first isolated from fatal cases of severe respiratory disease in horses and humans in 1994 in the Brisbane suburb of Hendra, Australia, and was shown to be distantly related to members of the *Morbillivirus* genus, resulting in its initial labelling as equine morbillivirus. Interestingly, there was a second and nearly concurrent spillover of HeV near Mackay, Queensland, ~1000 km north of Hendra, that was only retrospectively identified, where another person experienced a brief meningitic illness after assisting in the necropsies of two horses that where later shown to have died from HeV infection. This individual later succumbed to a fatal recrudescence of HeV-mediated meningoencephalitis, and in all there were 3 cases of human infection with two fatalities. NiV emerged a few years later in peninsular Malaysia in 1998-99, in a large outbreak of respiratory disease in pigs and encephalitis among pig farmers along with infection of abattoir workers in Singapore as a result of the transport of infected pigs. NiV was shown to be closely related both antigenically and genetically to HeV, and altogether, 265 cases and 105 deaths in people were reported in Malaysia with 11 cases and one fatality in Singapore (reviewed in [2]).

Both HeV and NiV have continued to cause spillover events from their natural bat reservoirs. In particular, HeV has appeared annually since 2006 and all through the recognition of infection of horses in Queensland and New South Wales, Australia, along with a total of 7 cases of human infection of which 4 have been fatal. In 2011 a significant increase in HeV spillovers began and 18 separate instances of infection in horses were recorded along with the first occurrence of infection in a dog, and as of September 2012 there have been 7 more episodes reported (reviewed in [8]). To date there has now been 39 HeV spillovers since 1994. Similarly, nearly annual outbreaks of NiV infection have been recorded since 2001, however these episodes have been primarily in Bangladesh but also including India (reviewed in [8]), with the most recent in January of 2012 involving 6 fatalities. The NiV outbreaks in Bangladesh have had significantly higher case fatality rates in people (typically between 75-100%) with several outbreaks linked to the consumption of raw date palm sap contaminated with NiV as well as human-to-human transmission of infection, and in total there have been 570 reported cases of NiV infection in people with 305 fatalities (reviewed in [8]).

Hendra Virus and Nipah Virus Host Range and Pathogenesis

The exceptionally broad species tropism of HeV and NiV are unique and in addition to bats, that do not experience clinical disease as a result of either natural or experimental henipavirus infection, the infection of horses, pigs, cats, dogs, ferrets, hamsters, guinea pigs, monkeys and humans will likely result in moderate to severe and often fatal disease. HeV and NiV will infect susceptible host cells by the coordinated actions of their attachment (G) and fusion (F) glycoproteins (reviewed in [9]). The F glycoprotein is a type I transmembrane glycoprotein and consists of two disulfide-linked subunits, F_1 and F_2, resulting from the proteolytic processing of the F_0 precursor. The F glycoprotein facilitates the membrane fusion step of the virus infection process by the actions of two α-helical heptad repeat domains known as (HR-1 and HR-2) that mediate the formation of a 6-helix bundle structure during or concomitant with membrane merger. The henipavirus attachment G glycoprotein is a type II transmembrane glycoprotein consisting of a stalk and globular head and the head structures have recently been determined (reviewed in [10]). Henipavirus infection tropism is determined by their utilization of the host cell membrane proteins ephrin-B2 and ephrin-B3 as receptors, and the binding of virus to cells is mediated by the G glycoproteins. These B-class ephrins are members of a large family of cell surface expressed ligands that bind to Eph receptor tyrosine kinases mediating important bi-directional cell-cell signalling events within the vascular, nervous and skeletal systems (reviewed in [10]).

The recognition of ephrin-B2 and -B3 as the host cell receptors for henipaviruses helped illuminate their broad species tropism and the systemic distribution of virus observed in infected hosts (reviewed in [3]). Indeed, the henipaviruses cause a systemic infection characterized by wide-spread vasculitis and endothelial cell tropism (reviewed in [3]). However, in addition to acute clinical disease, henipavirus infection in humans can take a more prolonged course following recovery from an initial infection, and the recrudescence of virus replication in the central nervous system (CNS) has frequently occurred, a process first noted in the second fatal case of HeV, but has been well-documented in numerous cases of human NiV infection with recrudescence of virus and CNS disease occurring from several months to as long as 11 years following infection [11]. How henipaviruses cause this apparent asymptomatic infection and recrudescence within the CNS are unknown and need further investigation.

Animal Modelling of Henipavirus Disease

The establishment and characterization of suitable animal models for henipavirus infection is essential for studying features of pathogenesis, the processes of virus shedding and transmission as well as the underlying mechanisms of recrudescence. In addition, animal models are required for the testing and evaluation of vaccines and therapeutics against henipavirus infection for use in humans and economically important livestock. Among recognized susceptible mammalian hosts, including humans, both HeV and NiV cause a systemic infection and the central features of acute henipavirus pathogenesis are essentially a wide-spread vasculitis with endothelial cell tropism with syncytia, thrombosis, ischemia and necrosis, along with parenchymal cell infection and pathogenesis of many major organs but is particularly prominent in the lung, brain, kidney and heart (reviewed in [3]). The major involvement of both the lung and brain correlate with severe henipavirus disease in humans that often manifests as acute interstitial pneumonitis, encephalitis or a combination of both.

Several, now well-established, animal models of HeV and NiV infection and pathogenesis have been reported and recently reviewed in detail elsewhere, and include the guinea pig, hamster, cat, pig, ferret, African green monkey (AGM) and horse [1]. Those models that most closely replicate henipavirus pathogenesis in humans are the hamster, ferret and AGM; whereas the most appropriate models for livestock vaccine development are those relevant species which are naturally infected in spill over events: the horse and pig. However a re-evaluation of HeV infection in the dog now appears warranted because of the recent naturally acquired infection reported in 2011 in that species. Here, brief descriptions of only the major pathogenic outcomes in those animals examined in the context of the henipavirus passive and active vaccination studies will be mentioned.

Passive Immunization Strategies

Passive immunization experiments were first carried out in hamster NiV-challenge models and provided critical proof-of-concept data showing that antibody-based immunotherapy against henipavirus infection was feasible. Protective passive immunization of hamsters was demonstrated with both NiV G and F glycoprotein specific polyclonal antiserums, as well as with mouse monoclonal antibodies (mAbs) also specific for the G or F glycoproteins [12, 13]. These studies also demonstrated the importance of henipavirus glycoprotein specific antibody in protection from disease (reviewed in [2]). Nevertheless, a successful and viable passive immunization therapy for use in people is likely to be one administered systemically and following a known exposure and/or infection by virus and also a renewable reagent such as a potent mAb that has either been humanized or is a human mAb.

Henipavirus-neutralizing human mAbs specific for the G glycoprotein have previously been isolated using recombinant antibody methods. Several mAbs were isolated from a large naïve human phage-displayed Fab fragment antibody library using purifed recombinant HeV soluble G glycoprotein (HeV-sG) as the panning antigen. From this panel of human mAbs one, termed m102, exhibited strong cross-reactive neutralization activity against live HeV and NiV and was affinity maturated followed by re-isolation of individual clones with improved affinity for HeV-sG leading to the development of m102.4 which was converted to IgG1 format and produced in a CHO-K1 cell line [14]. Human m102.4 mAb has exceptionally potent NiV and HeV neutralizing activity, with 50% inhibitory concentrations below 0.04 and 0.6 μg/ml respectively and its epitope maps to the ephrin receptor binding site (Fig. 1) [10]. Further, mAb m102.4 has been shown to neutralize all available isolates of HeV and NiV, including NiV-Malaysia, HeV-1994, HeV-Redlands and NiV-Bangladesh [15].

The protective efficacy of human mAb m102.4 was first tested in a lethal ferret model of NiV infection and to date is the only human mAb therapy for treating henipavirus infection that has been examined [15] (Table 1). In a pre- and post-exposure NiV challenge study in the ferret, groups of three animals each were given a single 50 mg dose (~25mg/kg) at 24 hrs before (pre-) or 10 hrs following (post-) challenge. The mAb m102.4 was administered by intravenous infusion with control ferrets given PBS, and the animals were challenged oronasally with a 5,000 $TCID_{50}$ of NiV (10-fold the minimal lethal dose-50%). Control ferrets became febrile with clinical illness by day 6, with severe disease by day 8 and were euthanized. By day 8, one animal in the pre-challenge group and all three in the post-challenge group were febrile with signs of depression, and fever began to abate by day 10. Two of 3 ferrets in the pre-challenge group later experienced disease progression and were euthanized at day 13, however, all other ferrets (3/3 in the post-challenge and 1/3 pre-challenge) were well and free of disease signs and remained well until the study end point (day 20). Upon study completion, the control animals revealed severe

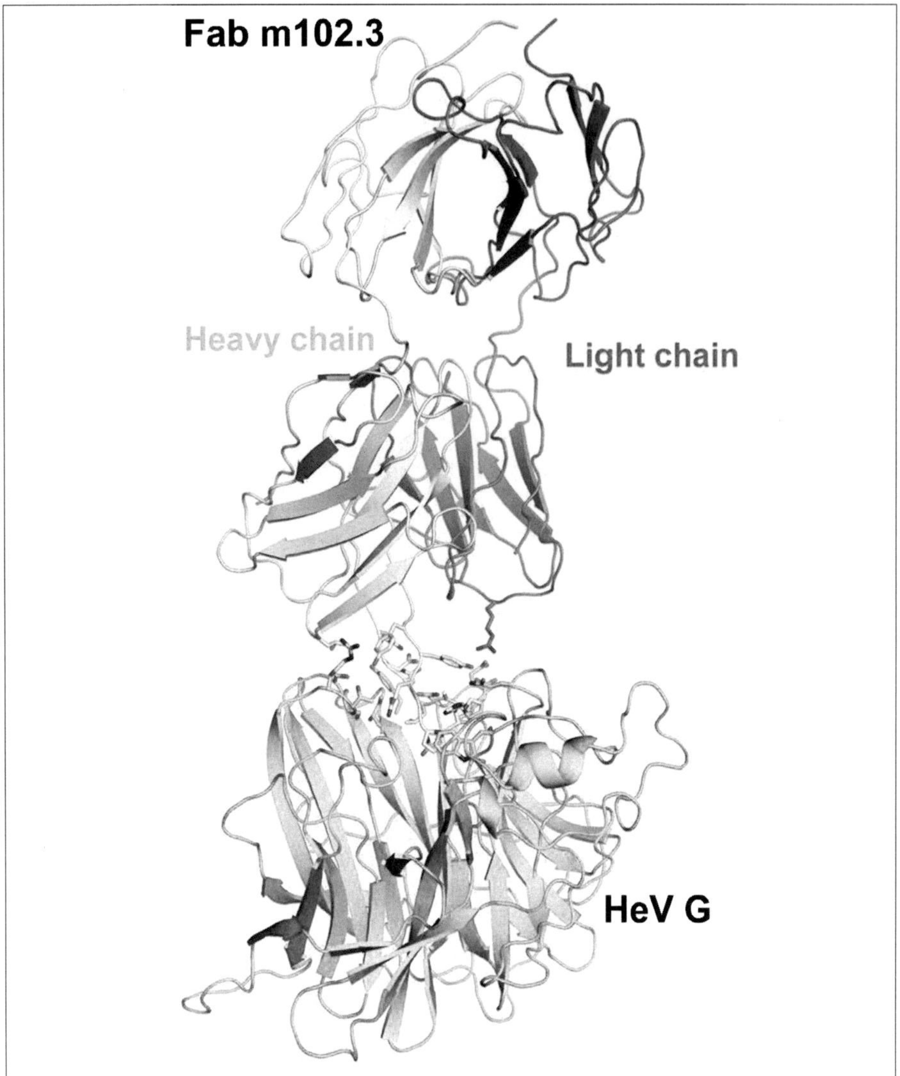

Fig. 1. Crystal structure of the Fab fragment of neutralizing monoclonal antibody (m102.3), a closely related mAb to m102.4 containing an identical heavy chain sequence, bound to the HeV G glycoprotein globular head. The tip of Fab's CDR (complimentary determining region)-H3 inserts in the same HeV-G hydrophobic cavity as the ephrin-B2 and –B3 G-H loop residues [10]

Table 1. Advanced passive immunization and active vaccination strategies evaluated in animal models by henipavirus challenge.

Strategy	Viral antigen target or immunogen	Challenge model
Passive immunization		
Human monoclonal antibody m102.4	Hendra / Nipah G glycoprotein	Ferret[a] (NiV)
		Nonhuman primate[b] (HeV)
Active vaccination		
Recombinant vaccinia virus	Nipah F and/or G glycoprotein	Hamster[c]
Recombinant canarypox virus	Nipah F and/or G glycoprotein	Pig[d] (NiV)
Subunit glycoprotein	Hendra soluble G glycoprotein	Cat[e] (NiV), Ferret[f] (HeV, NiV), nonhuman primate[g] (HeV, NiV) Horse[h] (HeV)

[a] A cross-reactive neutralizing human mAb against henipavirus G glycoprotein (m102.4) was shown to protect ferrets against lethal Nipah virus challenge by post-exposure infusion [15] and also Hendra virus (J. Pallister and C. Broder, unpublished).

[b] Human mAb m102.4 protects African green monkeys by post-exposure infusion following lethal Hendra virus challenge [16] and also Nipah virus challenge (T Geisbert and C. Broder, unpublished).

[c] NiV F and/or G encoding recombinant vaccinia viruses were used to immunize hamsters and protected against disease following intraperitoneal Nipah virus challenge [12].

[d] Nipah F and/or G glycoprotein encoding recombinant canarypox viruses are used to immunize pigs and protects against intranasal Nipah virus challenge [17]

[e] Hendra virus soluble G glycoprotein (HeV-sG) is used to immunize cats and protects against lethal subcutaneous Nipah virus challenge and against lethal oronasal Nipah virus challenge [18].

[f] HeV-sG is used to immunize ferrets and protects against lethal oronasal Hendra virus challenge [19]; and also Nipah virus challenge (Pallister and Broder, unpublished).

[g] HeV-sG is used to immunize African green monkeys and protects against lethal intratracheal Nipah virus challenge [20] and also Hendra virus challenge (Geisbert and Broder, unpublished).

[h] HeV-sG is used to immunize horses and protects against lethal oronasal Hendra virus challenge (D. Middleton and C. Broder, unpublished).

systemic pathology and the two pre-challenge treated animals had marked reduced frequency and size of hemorrhagic lesions in the pulmonary parenchyma. Notably, there were no pathological findings in any of the surviving ferrets including all in the post-challenge group that received a single dose of m102.4 10 hrs after lethal challenge [15]. Essentially identical outcomes with m102.4 therapy in a post-exposure study in the ferret following a HeV challenge have been generated (J. Pallister and C. Broder, unpublished).

Further pre-clinical development of m102.4 for future approved human use evaluation in a second animal model was required and the nonhuman primate, AGM, henipavirus challenge model employed [16] (Table 1). The study design for passive immunization in the AGM was designed to reflect a potential real life scenario requiring mAb as a post-exposure treatment. Here, fourteen monkeys were challenged intratracheally with a 10-fold lethal dose of HeV and 12 animals were infused twice with a 100 mg dose (~20mg/kg) of m102.4 beginning at 10 hr, 24 hr or 72 hr p.i. with the second dose given 48 hrs following the first in each group. All animals became infected following challenge and all 12 subjects that received m102.4 therapy survived infection while the untreated control subjects elaborated a severe systemic disease and succumbed 8 days following virus challenge. Subjects in the 72 hr post-exposure treatment group did exhibit disease signs but all these animals recovered by day 16. The mAb m102.4 had a distribution half-time (~1 day) and elimination half-time (~11 days) in the AGM. No evidence of HeV-specific pathology was observed in any of the m102.4-treated animals, no virus could be recovered and all tissues were negative for viral antigen, while controls had high levels of viral genome, recoverable virus in a wide-range of tissues along with significant viral antigen. This was the first successful post-exposure in vivo therapy against HeV and the first in a nonhuman primate. Essentially identical therapeutic outcomes with m102.4 in the AGM following a NiV challenge have been generated (T. Geisbert and C. Broder, unpublished).

Active Vaccination Strategies

The hamster challenge model was the first evaluation of an active vaccination strategy against NiV. The attenuated vaccinia virus strain, NYVAC, was used to generate NiV F and G glycoprotein gene encoding recombinant viruses and used to immunize animals, both individually and in combination, demonstrating that complete protection from NiV-mediated disease was achievable [12], however, NiV replication and an anamnestic antibody response in challenged animals was noted. The study confirmed that an immune response to the viral F and/G glycoproteins are important in protection.

The development of a henipavirus livestock vaccine has also been an important goal because of the repeated outbreaks of HeV in horses in Australia and also the association of pigs as an amplifying host for NiV. The first vaccines tested in this regard were recombinant canarypox-based (ALVAC) vectors encoding NiV F or G glycoprotein genes for possible use in swine [17] (Table 1). The ALVAC vectors were used to immunize 4-week-old pigs twice within 2 weeks, and each was tested alone and in combination. NiV neutralizing antibody was elicited by both vectors, with ALVAC-G yielding the strongest response. Piglets were challenged intranasally with 2.5 x 10^5 plaque-forming units (PFU) of NiV. The goals of the study in pigs were to demonstrate prevention of disease and virus shedding. Here, protection from NiV-mediated disease in all vaccinated pigs by either ALVAC vector alone or in combination was shown and only low levels of detectable virus shedding by nucleic acid analysis was reported, however no infectious virus could be recovered. These data suggested that either ALVAC vector could serve as a protective vaccine against NiV disease in pigs, but further studies may be needed to detail infectious virus shedding in these animals [17].

The most extensively examined henipavirus vaccine to date is a subunit immunogen consisting of a recombinant expressed, soluble and secreted, oligomeric form of the HeV G glycoprotein (HeV-sG); HeV-sG is now commercially available as an equine vaccine in Australia (Equivac® HeV, Pfizer Animal Health). The HeV-sG subunit vaccine (Fig. 2) is a version of the molecule that lacks the transmembrane and cytoplasmic tail domains and is produced in mammalian cell culture systems, is properly N-linked glycosylated, oligomeric and can bind ephrin-B2 and -B3 receptors (reviewed in [2]) .

HeV-sG can elicit a potent cross-reactive polyclonal neutralizing antibody response in mice, rabbits, cats, ferrets, monkeys and horses, and HeV-sG appears to present more cross-reactive epitopes (anti-NiV G reactive) as compared to sG from NiV. The henipavirus cat model was employed first in evaluating sG as a vaccine, with a side-by-side comparison made between sG from HeV and NiV. The sG immunogens were formulated in CSIRO triple adjuvant (Montanide, Quil A, and DEAE-dextran) and cats were given 3 doses of 100 μg at three week intervals. A completely protective immune response against a subsequent lethal subcutaneous NiV challenge was reported and also suggested that a single vaccine (HeV-sG) could be effective against both HeV and NiV (reviewed in [2]) (Table 1). A second study in the cat examined the degree of pre-challenge virus-neutralizing antibody titers required for protection and also employed a high dose oronasal challenge [18]. Here, only 2 doses of either 50 μg, 25 μg or 5 μg of HeV-sG formulated with CpG and Allhydrogel™ were given at day 0 and 21. Neutralizing antibody titers to NiV at day 35 ranged from 1:32 (low dose) to 1:512 (high dose), and cats were challenged on day 42. All vaccinated animals were completely protected from a high dose oronasal challenge of NiV (50,000TCID$_{50}$), while all controls succumbed to severe systemic disease.

A HeV-sG immunization study was also carried out in ferrets with a HeV challenge where animals were immunized twice (day 0 and day 20) with either a 100 μg, 20 μg or 4 μg dose of HeV-sG formulated in CpG and Allhydrogel™, and control subjects received adjuvant alone [19]. Again, a significant neutralizing antibody response was shown and correlated to the dose of HeV-sG used, with titers ranging from 1:8192 in the 100 μg group; 1:1024 and 1:2048 in the 20 μg; and 1:64 and 1:128 for the 4 μg. Ferrets were challenged oronasally with a 5,000 $TCID_{50}$ dose (10-fold the minimal lethal dose 50%) of HeV (Redlands 2008 isolate). All vaccinated ferrets remained completely free of any signs of fever or clinical disease from HeV infection, whereas the control ferrets developed fever, depression, and tremors, and extensive gross pathological lesions were noted. All HeV-sG vaccinated ferrets at the end of the study were gross pathologically and histopathologically normal except one of four animals in the lowest dose vaccine group. No evidence of virus antigen or genome was detected in any tissues or body fluids in any animals in the 100 μg and 20 μg vaccine groups; while only a low level of genome was detected in the nasal washes from one of four animals in the 4 μg (low dose) vaccine group. Further, no infectious virus was recovered from any vaccinated ferrets and together, these findings confirmed that subunit vaccination strategy against henipaviruses to prevent disease and virus shedding appeared possible. Essentially identical findings have also been noted in HeV-sG vaccinated ferrets challenged with NiV (J. Pallister and C. Broder, unpublished).

HeV-sG vaccination of nonhuman primates (AGMs) followed by intratracheal NiV challenge has recently been reported (Table 1) [20]. In this study, subjects were vaccinated with HeV-sG doses of 10, 50 or 100 μg formulated in CpG and Allhydrogel™ at day 0 and day 20, with a control subject receiving adjuvant alone, similar to the ferret study. Pre-challenge NiV serum neutralization titers ranged from 1:28 to 1:379. The high dose of HeV-sG (100 μg) appeared to generate the most sustained IgG response. Complete protection against subsequent NiV infection was observed in all vaccinated subjects with no evidence of clinical disease, virus replication, shedding or pathology seen in any challenged subjects, whereas the control subject developed severe systemic and fatal NiV disease, consistent with historical controls. Essentially identical findings have also been noted in HeV-sG vaccinated AMGs challenged with HeV (Geisbert T. and Broder C., unpublished).

Finally, the application of HeV-sG as a safe and effective equine vaccine has been pursued as a strategy to both prevent infection in horses and disrupt the risk of HeV transmission to people. A HeV-sG is now available for use in Australia. The HeV-sG based vaccine(Equivac® HeV) was commercially developed by Pfizer Animal Health as an equine vaccine for use in Australia. Horse HeV-sG vaccination and HeV challenge studies have recently been carried in Australia at the BSL-4 facilities of the Animal Health Laboratories (AAHL) in Geelong, Australia (reviewed in [8]). HeV-sG was used to immunize horses using 2 doses with a 3 week interval and a high and a low dose of HeV-sG antigen was examined, achieving pre-challenge HeV neutralizing titers between 1:128 to 1:2048. A high dose oronasal challenge with $2x10^6$ $TCID_{50}$ of HeV was employed and all HeV-sG vaccinated horses remained clinically disease-free. Further, there was no evidence of virus replication or virus shedding in any of the immunized horses. Equivac® HeV was launched in November 2012.

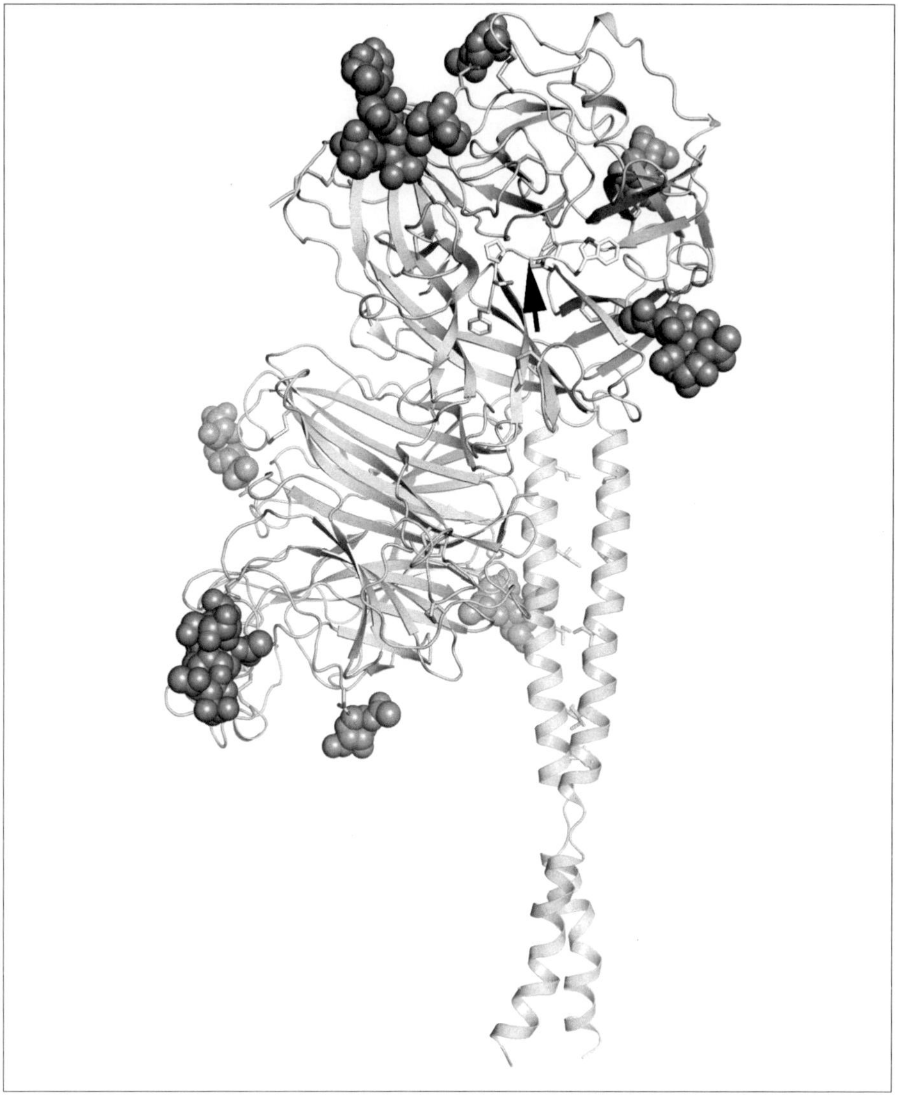

Fig. 2. The Hendra virus soluble G glycoprotein (HeV-sG) subunit vaccine. The HeV-sG glycoprotein subunit vaccine is composed of the entire predicted ectodomain (amino acids 76-604) and is shown as the dimer with two globular heads and two α-helical stalk domains. The secondary structure elements of the two globular head domains are derived from the crystal structure of the HeV G head domain, which also revealed that all five predicted N-linked glycosylation sites (N306, N378, N417, N481 and N529) were occupied by carbohydrate moieties (gray spheres). Here, N378 is not modeled in the figure due to weak electron density and the structures of the α-helical stalk domains, residues 77-136 of each monomer, are modeled revealing discontinuous helix. The ephrin receptor binding face of the globular head monomer on top (highest) is facing front and the amino acid side chains of the ephrin G-H loop residues are also shown and marked by the black arrow.

CONCLUDING REMARKS

HeV and NiV are the only known zoonotic paramyxoviruses that can infect and cause lethal disease across a broad range of vertebrate species including humans. Both viruses can also be isolated from a variety of bat reservoirs, and because of their associated high morbidity and mortality rates in humans and economically important livestock, they pose a significant biothreat that could occur as a result of natural outbreaks, laboratory accidents or their deliberate dispersal. A considerable amount of henipavirus research over the past decade has focused on the viral envelope glycoproteins that are critical for cell attachment and infection, and also the major targets of the host's neutralizing antibody response, making them obvious vaccine candidates (active and passive immunization). As a viable passive immunization approach for human use, a cross-reactive human mAb (m102.4), capable of neutralizing all known isolates of HeV and NiV, has been shown to be exceptionally efficacious as a post-exposure therapy in protecting both ferrets and nonhuman primates from lethal henipavirus disease, even under conditions of high dose virus challenge and by oronasal or intratracheal routes of virus inoculation. To date, m102.4 has been administered to three individuals in Australia on a compassionate use basis who were considered to be at high risk of HeV infection, and it is presently in pre-clinical development stages in both the United States and Australia.

As an active vaccination strategy against henipavirus infection, a recombinant subunit vaccine, HeV-sG, has shown remarkable cross-protective efficacy against both HeV and NiV. Complete protection against henipavirus challenge has been demonstrated in cats, ferrets, nonhuman primates and horses, with most studies showing that only two immunizing doses can elicit a protective antibody response even from high dose virus challenge, along with no evidence of virus shedding in immunized subjects. The HeV-sG subunit vaccine (Equivac® HeV, Pfizer Animal Health) is now commercially available as an equine vaccine for use in Australia. Findings to date suggest the HeV-sG subunit vaccine may also be suitable for use in humans and could be beneficial for administration to veterinarians and staff workers working with horses in Australia, researchers working with HeV or NiV, or people at risk of NiV infection in Bangladesh.

ACKNOWLEDGEMENTS

The views expressed in the manuscript are solely those of the author, and they do not represent official views or opinions of the Department of Defense or The Uniformed Services University. Special thanks to Kai Xu of the Structural Biology Program, Memorial Sloan-Kettering Cancer Center, New York for developing the structural model of the HeV-sG vaccine and m102.3 Fab complex with HeV-sG. This work was supported in part by the Department of Health and Human Services, National Institutes of Health, grants AI054715 and AI077995.

Conflict of Interest

The author is a United States federal employee and coinventor on patents relating to human monoclonal antibodies against Hendra and Nipah viruses, soluble forms of Hendra and Nipah envelope glycoproteins and vaccines; assignees are The United States of America as represented by the Department of Health and Human Services (Washington, DC), Henry M. Jackson Foundation for the Advancement of Military Medicine, Inc. (Bethesda, MD).

References

1 Geisbert TW, Feldmann H, Broder CC: Animal challenge models of Henipavirus infection and pathogenesis. Curr Top Microbiol Immunol 2012;359:153-77.

2 Broder CC: Therapeutics and vaccines against Hendra and Nipah viruses; in Levine MM, Dougan G, Good MF, Liu MA, Nabel GJ, Nataro JP, et al., editors. New Generation Vaccines. 4 ed. New York: Informa Healthcare USA; 2010. p. 885-94.

3 Wong KT, Ong KC: Pathology of acute henipavirus infection in humans and animals. Patholog Res Int 2011;2011:567248.

4 Pernet O, Wang YE, Lee B: Henipavirus receptor usage and tropism. Curr Top Microbiol Immunol 2012;359:59-78.

5 Aguilar HC, Iorio RM: Henipavirus membrane fusion and viral entry. Curr Top Microbiol Immunol 2012;359:79-94.

6 Middleton DJ, Weingartl HM: Henipaviruses in their natural animal hosts. Curr Top Microbiol Immunol 2012;359:105-121.

7 Clayton BA, Wang LF, Marsh GA: Henipaviruses: An updated review focusing on the pteropid reservoir and features of transmission. Zoonoses Public Health 2012; doi: 10.1111/j.1863-2378.2012.01501.x.

8 Broder CC, Geisbert TW, Xu K, Nikolov DB, Wang LF, Middleton D, et al: Immunization strategies against Henipaviruses. Curr Top Microbiol Immunol 2012;359:197-223.

9 Steffen DL, Xu K, Nikolov DB, Broder CC: Henipavirus mediated membrane fusion, virus entry and targeted therapeutics. Viruses 2012;4 doi:(2):280-308.

10 Xu K, Broder CC, Nikolov DB: Ephrin-B2 and ephrin-B3 as functional henipavirus receptors. Semin Cell Dev Biol 2012;23(1):116-23.

11 Abdullah S, Chang LY, Rahmat K, Goh KT, Tan CT: Late-onset Nipah virus encephalitis 11 years after the initial outbreak: A case report. Neurology Asia 2012;17(1):71-174.

12 Guillaume V, Contamin H, Loth P, Georges-Courbot MC, Lefeuvre A, Marianneau P, et al: Nipah virus: vaccination and passive protection studies in a hamster model. J Virol 2004;78(2):834-840.

13 Guillaume V, Wong KT, Looi RY, Georges-Courbot MC, Barrot L, Buckland R, et al: Acute Hendra virus infection: Analysis of the pathogenesis and passive antibody protection in the hamster model. Virology 2009;387(2):459-465.

14 Zhu Z, Bossart KN, Bishop KA, Crameri G, Dimitrov AS, McEachern JA, et al: Exceptionally potent cross-reactive neutralization of Nipah and Hendra viruses by a human monoclonal antibody. J Infect Dis 2008;197(6):846-853.

15 Bossart KN, Zhu Z, Middleton D, Klippel J, Crameri G, Bingham J, et al: A neutralizing human monoclonal antibody protects against lethal disease in a new ferret model of acute nipah virus infection. PLoS Pathog 2009;5(10):e100064.

16 Bossart KN, Geisbert TW, Feldmann H, Zhu Z, Feldmann F, Geisbert JB, et al: A neutralizing human monoclonal antibody protects African green monkeys from hendra virus challenge. Sci Transl Med 2011;3(105):105ra3.

17 Weingartl HM, Berhane Y, Caswell JL, Loosmore S, Audonnet JC, Roth JA, et al: Recombinant nipah virus vaccines protect pigs against challenge. J Virol 2006;80(16):7929-7938.

18 McEachern JA, Bingham J, Crameri G, Green DJ, Hancock TJ, Middleton D, et al: A recombinant subunit vaccine formulation protects against lethal Nipah virus challenge in cats. Vaccine 2008;26(31):3842-3852.
19 Pallister J, Middleton D, Wang LF, Klein R, Haining J, Robinson R, et al: A recombinant Hendra virus G glycoprotein-based subunit vaccine protects ferrets from lethal Hendra virus challenge. Vaccine 2011;29(34):5623-5630.
20 Bossart KN, Rockx B, Feldmann F, Brining D, Scott D, Lacasse R, et al: A Hendra virus G glycoprotein subunit vaccine protects African green monkeys from nipah virus challenge. Sci Transl Med 2012;4(146):146ra07.

Christopher C. Broder, PhD, Department of Microbiology and Immunology, Uniformed Services University, Bethesda, MD 20814, USA
Tel: 301-295-3401, fax: 301-295-1545
Email: christopher.broder@usuhs.edu

Roth JA, Richt JA, Morozov IA (eds): Vaccines and Diagnostics for Transboundary Animal Diseases. Dev Biol (Basel). Basel, Karger, 2013, vol 135, pp 139-145.

Current Status of Diagnostic Methods for Henipavirus

A. Tamin, P. A. Rota

Division of Viral Diseases, Centers for Disease Control and Prevention, Atlanta, GA, USA

Key words: *Henipavirus*, diagnostic, Nipah virus, Hendra virus, zoonotic diseases

Abstract: Hendra virus (HeV) and Nipah virus (NiV) are the causative agents of emerging transboundary animal disease in pigs and horses. They also cause fatal disease in humans. NiV has a case fatality rate of 40 – 100%. In the initial NiV outbreak in Malaysia in 1999, about 1.1 million pigs had to be culled. The economic impact was estimated to be approximately US$450 million. Worldwide, HeV has caused more than 60 deaths in horses with 7 human cases and 4 deaths. Since the initial outbreak, HeV spillovers from *Pteropus* bats to horses and humans continue. This article presents a brief review on the currently available diagnostic methods for henipavirus infections, including advances achieved since the initial outbreak, and a gap analysis of areas needing improvement.

Introduction

Hendra virus (HeV) and Nipah virus (NiV) are members of genus *Henipavirus* within the virus family *Paramyxoviridae*. Both cause deadly transboundary zoonotic diseases. Both infect and cause severe disease in a number of host species. Due to the severity of the disease and the lack of vaccines or anti-viral therapeutics, NiV and HeV are classified as BSL-4 agents. Their reservoir is bats of the genus *Pteropus*. Further details of the molecular virology, epidemiology, pathogenesis and natural history of *Henipaviruses* can be obtained from published reviews [5, 14].

HeV was identified in Australia in 1994 [5, 19]. Since then, sporadic spillovers of HeV continue, causing deaths of horses, and 7 human cases with 4 deaths from close contacts with infected horses [16]. No human-to-human transmission of HeV has been reported.

NiV emerged in Malaysia in 1999, resulted in 265 human cases with 105 deaths [5]. To control the outbreak, 1.1 million pigs were culled. The Office International des Epizooties (OIE) declared the Malaysia pig population officially free of NiV infection in 2001 [1]. In 2003, NiV emerged in Bangladesh and India. Human-to-human transmission occurred without an amplifying animal host; and consumption of contaminated palm juice was often identified as the source of human infection. The human toll of NiV in Bangladesh was 278 cases with 194 deaths [14].

The purpose of this brief review is to summarize the progress and current status of diagnostic methods for henipaviruses, and to identify areas for further research and development. More details about each method presented can be found in recent reviews [15, 19].

Clinical Diagnosis

The clinical presentation of Nipah virus infection in animals is sometimes used for diagnosis during outbreaks, but it is not specific [1]. A recent review describes the clinical features of human henipavirus infection, but laboratory tests including pathological examinations are required to confirm the clinical diagnosis [20].

Laboratory Diagnosis

Virus isolation and electron microscopy

Virus isolation in Vero cells or recently available pteropid bats cells [3] is important and desirable, but requires BSL-4 containment and has a very slow turnaround time. Although syncytia formation in cell culture is the characteristic cytopathic effect (CPE) of *henipavirus*, it is not specific because many other viruses can cause syncytia formation. Therefore, another test to identify the virus is required.

Electron microscopy (EM)

Direct electron microscopic observation continues to be an indispensable of pathogen identification. However, it has limited utility during outbreaks when the etiologic agent has been identified [9].

Immunohistochemistry (IHC)

IHC is a useful method to study pathogenesis and for pathologic investigations of HeV and NiV infections by visualizing antigen-antibody complexes in the tissue samples. IHC offers an effective method to detect henipavirus from archival formalin-fixed samples. The strengths and weaknesses of IHC have been described previously [8].

Serology - Immunofluorescence assay (IFA)

IFA to detect specific antibody, antigen or virus offers a simple, convenient and quick test, but is not quantitative. Monospecific and/or monoclonal antibody against N, P, W, or C proteins allow IFA to study their intracellular distribution [12].

Serology - Enzyme-linked immunosorbent assay (ELISA)

ELISA is perhaps the most widely used serologic test. It is cost-effective, convenient, rapid, quantitative, and can be automated. ELISA can be configured to detect viral antigen or virus-specific IgG and IgM antibodies. The presence of IgM indicates a current, primary infection, while an IgG can be used to detect a past exposure or infection. Antibody detection is more widely used since it can be safely performed at BSL-2. During the NiV outbreak in Malaysia, antigens prepared from henipavirus-infected Vero cells (prepared in BSL-4 containment and gamma-irradiated), were used to detect IgG (indirect-EIA) and IgM (capture-EIA). Today a variety of ELISA antigens, including those prepared entirely in BSL-2, are available, such as henipavirus N, M, P and G proteins [19].

Serology - Neutralization assays and its surrogates

The serum neutralization test (SNT), which uses live virus and measures functional antibody, is considered as the gold standard for detecting IgG to henipaviruses. However, SNT requires BSL-4 containment and has limited availability. To overcome this limitation, various alternative approaches to SNT have recently been developed. First, a modified SNT, rapid immune plaque assay, offers better turnaround as well as less BSL-4 containment time [4]. In this test, infection or degree of inhibition of infection is read at BSL-2 by detecting and enumerating foci of infection by immunostaining on the decontaminated plates.

Luminex-based binding or blocking test uses the surface G proteins (sG) of HeV and NiV that are coated on Luminex beads. Antibodies bound to the coated beads are subsequently detected using biotinylated Protein-A/G chemistry in a Bio-Plex Protein Array System (BioRad, USA). In the blocking test, the same sG-coated Luminex beads are used; but the Protein-A/G mix is replaced by a biotinylated soluble ephrin-B2 molecule. The presence of henipavirus specific antibodies will block the binding between biotin ephrin-B2 and sG proteins on the beads. The Luminex blocking assay can be safely performed in a standard laboratory without any bio-containment need [2]. This assay was successfully used in confirming human HeV infections, for sero-surveillance in bats in Africa and to detect HeV infection in dogs [19].

In the vesicular stomatitis virus pseudotype system (pVSV), the VSV attachment protein G is replaced with luciferase [18] or green fluorescent protein (GFP) as a reporter system [11]. The NiV F and G proteins are expressed on the surface of the VSV pseudotype by plasmid expression vectors supplied *in-trans*. The neutralization assays using VSV pseudotypes as antigen is similar to the traditional SNT, but the pseudotype assay can be performed at BSL-2, and the results can be measured in 48 hours. A second generation of pseudotyped virus was recently developed, using secreted alkaline phosphatase (SEAP) instead of GFP or luciferase, so that results can be read on an ELISA plate reader [10]. The pVSV-NiV-F/G particles could also serve as ELISA antigens [18].

Molecular – RT-PCR

Because of its high sensitivity, specificity, convenience and rapid turn-around time, molecular tests have become the diagnostic method of choice. Details of the utility of molecular methods to detect henipaviruses can be obtained from a previously published review [15]. However, it is important to note that contamination can affect PCR results; and specimen quality is very important. Negative PCR results do not rule out cases.

The first molecular test was a conventional, end point RT-PCR. Two target genes were used, the matrix (M) protein identified in the CSIRO Australian Animal Health Laboratory (AAHL) assay, and the nucleocapsid (N) protein identified in the US Centers for Disease Control and Prevention (CDC) assay [19].

Conventional PCR has been replaced by real-time RT-PCR (qPCR) in either the TaqMan or SYBR green formats. A TaqMan qPCR assay targeting HeV M gene was shown to be highly specific, and 1,000 times more sensitive than conventional PCR [17]. TaqMan qPCR targeting NiV N gene showed sensitivity to detect 1 plaque forming unit (PFU), and was highly NiV specific [7]. Feldman and co-workers evaluated different formats of qPCR with N, M and P gene targets as well as heminested PCR targeting L gene, and observed different sensitivities and specificities relative to format (TaqMan or SYBR green) and target genes used [6].

Molecular – sequencing

Sequencing played a crucial role in the initial discovery and characterization of HeV and NiV, and the designation of new genus, *Henipavirus*. Sequencing is also a very useful tool for genotyping and tracking virus transmission [13]. Furthermore, sequencing has a vital role for supporting molecular tests, by detecting nucleotide substitutions in target genes which may affect the sensitivity and specificity of the PCR primers or probes. In the future, more powerful next generation sequencing (NGS) methods will be applied for whole genome sequencing and characterizing new henipaviruses [19].

Conclusions

Sensitive and specific diagnostic tests are necessary to detect the causes and map the spread of outbreaks. These methods are also required to evaluate the effectiveness of vaccine and antiviral treatments (reviewed in this volume by Dr. Broder). Therefore, diagnostic assays play a vital role in disease control and prevention, and impact public health and economic activities. A major challenge for HeV and NiV diagnostics is that the viruses are listed as select agents as well as

BSL-4 infectious agents. Consequently, diagnostic testing comes with significant logistic and financial challenges. However, because of advances in molecular biology, digital technology and bioinformatics, significant progress in henipavirus diagnostics has been achieved, particularly in the serologic and molecular approaches (Table 1).

Table 1. Progress of different diagnostic methods for henipavirus

Diagnostic methods	Discovery and initial outbreak[a]	The second decade after initial outbreak[a]
ELISA	Gamma-irradiated HeV or NiV infected Vero cells lysate as ELISA antigen (prepared in BSL-4 containment)	Variety of recombinant DNA expression systems – expressing nucleocapsid (N), phosphoprotein (P), or attachment glycoprotein (G) proteins of HeV or NiV as ELISA antigen (prepared in BSL-2 containment)
Electron microscope	Limited to polyclonal antibody from convalescent patients or animals for immunogold labelling	Monoclonal and monospecific polyclonal antisera against specific proteins of HeV or NiV are available for immunogold labelling
Immunohisto-chemistry	Using polyclonal antibody from convalescent patients or animals	Monoclonal and monospecific polyclonal antisera against specific proteins of HeV or NiV are available
Virus neutralization assay	Serum neutralization test (SNT) using infectious virus (performed in BSL-4 containment)	Various formats of surrogate approaches to SNT: Luminex-based binding & blocking assay, vesicular stomatitis virus (VSV) pseudotype vector with either luciferase, GFP or secreted alkaline phosphatase (SEAP) as reporter system (performed entirely in BSL-2 containment)
Molecular (polymerase chain reaction)	Semi-quantitative conventional PCR or end-point PCR	Quantitative PCR (qPCR), either TaqMan or SYBR Green format (rapid, highly sensitive and specific)
Molecular (sequencing)	Sanger dideoxy sequencing (i.e. took many months to complete the whole genome sequence of HeV and NiV)	Sanger dideoxy sequencing for point mutations study and genotyping; and next generation sequencing (NGS) for rapid whole genome sequencing and advanced bioinformatics

[a] 1994 for HeV outbreak in Australia; 1999 for NiV outbreak in Malaysia and Singapore

From the perspective of a gap analysis, there are a few improvements for henipavirus diagnostics that still need to be considered. Because of the growing interest in using a HeV sG based vaccine to immunize animals such as horses to prevent spillover to humans, there is a need to develop validated diagnostic approaches that can differentiate infected and vaccinated animals (DIVA).

ELISA is perhaps the most widely used serologic test in most laboratories. Although we have a collection of different ELISA antigens that are produced entirely at BSL-2 containment, most of these ELISA formats and the corresponding antigens used still need to be evaluated and validated for their sensitivity and specificity. For example, some species of bats show a very high background when using baculovirus expressed NiV N as an indirect ELISA antigen (unpublished observation).

Henipavirus outbreaks often occur on farms, if not remote villages, far from diagnostic laboratories. Similarly, viral surveillance in wildlife, including bats and other animals, is often necessary in isolated areas. A deliberate release of HeV or NiV must be detected quickly. In these cases, rapid, on-site detection is necessary. Therefore, there is a need to develop a rapid field test that is cost-effective, yet sensitive and specific. Given the unpredictability of outbreaks caused by henipavirus it would also be prudent to collaboratively develop diagnostic capacity and to strengthen surveillance, especially in those countries where spillover infections from wildlife are likely to occur.

HeV and NiV are RNA viruses which have an increased capacity for genetic change. Such genetic changes may affect pathogenicity, host tropism or transmissibility. Therefore, it is important to use molecular epidemiologic methods to study transmission patterns during henipavirus outbreaks and to perform genomic sequencing on representative samples.

In conclusion, although none of the described diagnostic methods for henipavirus are available commercially, significant progress has been made in developing methods that can be performed under BSL-2 conditions, or in a BSL-2 laboratory. A greater variety of tests, including those that must be performed at BSL-4 conditions, are available in a few institutions such as AAHL and CDC. Therefore, sustained preemptive global collaboration and communication and reagent sharing is essential. Recent successes in mitigating SARS-CoV and Schmallenberg virus outbreaks through extraordinary international multidisciplinary collaboration does offer an excellent model and provide hope for facing future outbreaks of henipavirus.

References

1 FAO: Manual on the diagnosis of Nipah virus infection in animals RAP publication no. 2002/01. Food and Agricultural Organization (FAO) of the United nations; 2002.

2 Bossart KN, McEachern JA, Hickey AC, Choudhry V, Dimitrov DS, Eaton BK, Wang LF: Neutralization assays for differential henipavirus serology using Bio-Plex protein array systems. J Virol Meth 2007;142:29-40.

3 Crameri G, Todd S, Grimley S, McEachern JA, Marsh GA, Smith C, et al: Establishment, immortalisation and characterisation of pteropid bat cell lines. PloS One2009;4e8266.

4 Crameri G, Wang LF, Morrissy LF, White J, Eaton BT: A rapid immune plaque assay for the detection of Hendra and Nipah viruses and anti-virus antibodies. J Virol Meth. 2002;99:41-51.

5 Eaton B, Mackenzie J, Wang LF: Henipaviruses. Philadelphia, Lippincott Williams & Wilkins.

6 Feldman KS, Foord A, Heine HG, Smith IL, Boyd V, Marsh GA, et al: Design and evaluation of consensus PCR assays for henipaviruses. J Virol Methods 2009;161:52-57.

7 Guillaume V, Lefeuvre A, Faure C, MarianneauP, Buckland R, Lam SK, et al: Specific detection of Nipah virus using real-time RT-PCR (TaqMan). J Virol Meth;2004;120:229-237.

8 Hooper PT, Russell GM, Selleck PW, Lunt RA, Morrissy CJ, Braun MA, Williamson MM: Immunohistochemistry in the identification of a number of new diseases in Australia. Vet Microbiol 1999;68:89-93.

9 Hyatt AD, Zaki SR, Goldsmith CS, Wise TG, Hengstberger SG: Ultrastructure of Hendra virus and Nipah virus within cultured cells and host animals. Microbes and infection 2001;3:297-306.

10 Kaku Y, Noguchi A, Marsh GA, Barr JA, Okutani A, Hotta K, et al: Second generation of pseudotype-based serum neutralization assay for Nipah virus antibodies: sensitive and high-throughput analysis utilizing secreted alkaline phosphatase. J Virol Meth 2012;179:226-232.

11 Kaku Y, Noguchi A, Marsh GA, McEachern JA, Okutani A, Hotta K, et al: A neutralization test for specific detection of Nipah virus antibodies using pseudotyped vesicular stomatitis virus expressing green fluorescent protein. J Virol Meth 2009;160:7-13.

12 Lo MK, Harcourt BH, Mungall BA, Tamin A, Peeples ME, Bellini WJ, Rota PA: Determination of the henipavirus phosphoprotein gene mRNA editing frequencies and detection of the C, V and W proteins of Nipah virus in virus-infected cells. J Gen Virol 2009;90:398-404.

13 Lo MK, Lowe L, Hummel KB, Sazzad HM, Gurley ES, Hossain MJ, et al: Characterization of Nipah virus from outbreaks in Bangladesh, 2008-2010. Emerg Infect Dis 2012;18:248-255.

14 Luby SP, Gurley ES: Epidemiology of henipavirus disease in humans. Curr Top Microbiol Immunol 2012;359:25-40.

15 Rota P, Mungal B, Halpin K: Molecular Detection of Human Viral Pathogens, Ed Liu D, Chapter 49. Nipah virus. London, UK. CRC Press.

16 Smith I, Broos A, de Jong C, Zeddeman A, Smith C, Smith G, et al: Identifying Hendra virus diversity in pteropid bats. PloS One 2011;6:e25275.

17 Smith IL, Halpin K, Warrilow D, Smith GA: Development of a fluorogenic RT-PCR assay (TaqMan) for the detection of Hendra virus. J Virol Meth 2001;98:33-40.

18 Tamin A, Harcourt BH, Lo MK, Roth JA, Wolf MC, Lee B, et al: Development of a neutralization assay for Nipah virus using pseudotype particles. J Virol Methods 2009;160:1-6.

19 Wang LF, Daniels P: Diagnosis of henipavirus infection: current capabilities and future directions. Curr Top Microbiol Immunol 2012;359:179-196.

20 Wong KT, Tan CT: Clinical and pathological manifestations of human henipavirus infection. Curr Top Microbiol Immunol 2012;359:95-104.

Azaibi Tamin, Measles, Mumps, Rubella and Herpes Viruses Branch, MS C-22, Division of Viral Diseases, National Center for Immunization and Respiratory Diseases, Centers for Disease Control and Prevention, Atlanta, GA 30333, USA

Email: atamin@cdc.gov

Roth JA, Richt JA, Morozov IA (eds): Vaccines and Diagnostics for Transboundary Animal Diseases. Dev Biol (Basel). Basel, Karger, 2013, vol 135, pp 147-157.

Prospects for Development of African Swine Fever Virus Vaccines

L.K. Dixon, C.C. Abrams, D.D.G. Chapman, L.C. Goatley, C.L. Netherton, G. Taylor, H.H. Takamatsu

The Pirbright Institute, Ash Road, Pirbright, UK

Key words: African swine fever virus, vaccine, virulence genes, immune response

Abstract: African swine fever virus is a large DNA virus which can cause an acute haemorrhagic fever in pigs resulting in high mortality. No vaccine is available, limiting options for control. The virus encodes up to 165 genes and virus particles are multi-layered and contain more than 50 proteins. Pigs immunised with natural low virulence isolates or attenuated viruses produced by passage in tissue culture and by targeted gene deletions can be protected against challenge with virulent viruses. CD8+ cells are required for protection induced by attenuated strain OURT88/3. Passive transfer of antibodies from immune to naïve pigs can also induce protection. Knowledge of the genome sequences of attenuated and virulent strains and targeted gene deletions from virulent strains have identified a number of virus genes involved in virulence and immune evasion. This information can be used to produce rationally attenuated vaccine strains. Virus antigens that are targets for neutralising antibodies have been identified and immunisation with these recombinant proteins has been shown to induce partial protection. However knowledge of antigens which encode the dominant protective epitopes recognised by CD8+ T cells is lacking.

The African Swine Fever Virus

African swine fever virus (ASFV) causes a haemorrhagic fever in domestic pigs which can result in very high mortality rates. The virus is a large, icosahedral, double-stranded DNA virus that replicates in the cytoplasm of infected cells and has a tropism for macrophages. It is the only member of the *Asfarviridae* family [1].

Genome

The ASFV genome is between 170 and 193 kbp long, depending on the isolate and encodes between 150 and 165 open reading frames (ORFs). These ORFs encode proteins required for virus replication in cells, including enzymes required for replication and transcription of the genome and structural proteins. In addition the virus encodes many proteins that are not essential for replication but have a role in virus host interactions including proteins involved in evading the host immune response.

Complete genome sequences from 12 virus isolates have been published [2, 3]. Length variation between isolates is due to gain or loss of genes mainly within the left terminal 40 kbp and right terminal 20 kbp. Most variation is due to gain or loss of members of 5 different multigene families (MGFs) encoded by the virus. Members of the MGF 360 and 530 families have been shown to have roles in host cell tropism and suppression of the interferon response [4].

Molecular epidemiology

Virus isolates have been classified into different genotypes according to the partial sequence of the gene encoding the p72 major capsid protein. This has identified 22 different genotypes to date. These are all present in Eastern and Southern Africa where the ancient sylvatic cycle in warthogs and the soft tick vector *Ornithodoros moubata* is present. Genetic diversity in this wildlife cycle is very wide and spillover of genotypes to domestic pigs has been more restricted. For example a single genotype, I, was for many years the only genotype identified in Central and Western Africa. This genotype spread to Europe and to the Caribbean and South America in the late twentieth century and remains endemic in Sardinia [5-7]. More recently genotype IX has been identified in the Republic of Congo. This is the predominant strain circulating in domestic pigs in Uganda and Kenya and it is likely that this strain has spread from East Africa [8, 9]. Predominant genotypes circulating in domestic pigs in some African countries have been characterised but overall there is a lack of knowledge about the prevalence of ASF and the predominant genotypes circulating. In 2007 a genotype II ASFV strain was introduced into Georgia. This genotype had previously been identified in pigs in Mozambique, Zambia and Madagascar [2, 10]. ASF has since spread to countries neighbouring Georgia and has been reported from different regions in the Russian Federation. Disease is now considered endemic in southern Russia and increasing outbreaks have been reported from further north and west including a first appearance in Ukraine (OIE WAHID).

Phylogeny based on 125 conserved genes of the 12 complete genome sequences has identified 2 main virus clades, one comprising genotypes I, II and IV and the other more diverse clade contains other genotypes from East Africa [2, 3, 10].

Virus structure

Virus particles have a complex multi-layered structure consisting of a nucleoprotein core which contains the virus genome and enzymes required to initiate the next round of infection. This is surrounded by a core shell and an internal lipid membrane onto which the icosahedral capsid is assembled. Extracellular virions gain an external envelope as they bud through the plasma membrane. Virions contain more than 50 proteins. The localisation of a number of proteins has been mapped including the major core shell proteins which are derived by proteolytic processing from a 220 kDa polyprotein. The protein p72 (also known as p73) encoded by B646L gene is the major capsid protein. The B438L protein is required to form the vertices of the icosahedral capsid [11]. Proteins inserted into the inner membrane and which may be partially exposed on the surface include p54 (E183L) and p17 (D117), proteins on the external virus envelope include CD2v (EP402R), p22 (K117L) and p12 (O61R).

Replication

Macrophages are the main target cells for ASFV replication. Macrophages with cell surface markers characteristic of early and late stages of differentiation are permissive for replication [12, 13]. Most studies on virus entry have been carried out with tissue culture cells. Early data identified receptor-mediated endocytosis as the main entry mechanism and more recently entry via clathrin coated vesicles was identified as the principle endocytic pathway utilised. Entry via macropinocytosis has also been demonstrated. The scavenger receptor CD163, was suggested to act as receptor [13]. However, ASFV can replicate in cells that do not express CD163, suggesting that the virus can bind to other receptors. Since both the intracellular mature ASFV virions and extracellular enveloped virions are infectious, possibly the virus has more than one entry mechanism.

Virus genes are expressed in a cascade including immediate early, early, intermediate and late gene classes. Genome replication begins at about 6 hours post-infection. Virion assembly takes place in the cytoplasm in perinuclear factory areas. The mature virions move to the plasma membrane and bud through the plasma membrane gaining an additional membrane in the process [1, 14, 15].

Immunity to ASFV

It has been known for many years that pigs which recover from infection with ASFV are protected against challenge with closely related isolates [16]. Thus the prospects for development of effective vaccines are very good. Soon after the introduction of ASF to Portugal and Spain in the 1960s, attenuated strains were derived by passage of field isolates in cell culture [17]. These attenuated strains were then used in the field in Portugal and Spain but with disastrous consequences since many pigs developed adverse post-vaccination effects. Following this experience, there have been no further field trials of attenuated vaccine strains. Attempts to develop a killed vaccine were also unsuccessful.

Role of antibody in protection

Experimental vaccination has been used to identify mechanisms of protective immunity induced by attenuated ASFV strains. Early reports suggested that protective neutralizing antibodies were not produced during ASFV infection [18, 19]. However, some studies have demonstrated that neutralizing antibodies can play a role in protection. One study demonstrated that convalescent serum from pigs infected with a low virulence Spanish isolate, E75, neutralized the infectivity of virulent ASFV isolates and low passage tissue culture adapted variants, by 86-97% in Vero and macrophage cell cultures [20, 21]. A monoclonal antibody, mAb-135D4, recognizing a conformational epitope of an ASFV protein of 72 kDa also exhibited strong neutralizing activity with these viruses [22]. Unexpectedly, although mAb-135D4 reacted with all virulent viruses, neither E75 immune sera nor mAb-135D4 neutralized high passage tissue culture adapted ASFV variants, suggesting that tissue culture adaptation of ASFV isolates may be associated with loss of virus neutralization determinants [20], perhaps dependent on the phospholipid composition of the virus particles [23].

A key observation suggesting a role for antibodies in protection was the demonstration of protection of pigs against lethal infection through passive transfer of anti-ASFV serum [24]. A significant 3 day delay in onset of viraemia and a 10,000-fold reduction in both mean and maximum titres of virus in blood were observed in animals given anti-ASFV Ig. In addition to virus neutralization, a number of other mechanisms by which antibodies to ASFV may mediate protection have been suggested. Thus, antibodies that inhibit ASFV replication post-virus adsorption correlate with protection; complement-dependent antibody-mediated cytotoxicity and antibody-dependent cell-mediated cytotoxicity may also play a role. However, antibodies that inhibit neutralisation have been identified and these may play a role in reducing the efficacy of antibody-mediated protection [25, 26].

Role of cellular immunity in protection

An important role for CD8+ T cells in protection against ASFV was shown in studies in which protection induced by the attenuated ASFV strain OURT88/3 was abrogated by depletion of CD8+ cells using monoclonal antibodies [27]. ASFV-specific cytotoxic T lymphocytes (CTL) have been demonstrated in inbred miniature swine experimentally infected with a naturally occurring, non-fatal ASFV isolate (NHV) [28, 29]. Lysis of AFV-infected cells was shown to be mediated by purified CD8+ lymphocytes and blocked by anti-MHC class I monoclonal antibodies. Evidence for cytotoxic T lymphocytes was obtained, in a protection model in which syngeneic inbred miniature swine were experimentally inoculated with a naturally occurring, non-fatal ASFV isolate. The specific lysis was shown to be mediated by purified CD8+ lymphocytes and blocked by anti-class I monoclonal antibodies. In addition, purified lymphocytes from pigs immunised with attenuated strains produce high levels of interferon-gamma after stimulation with ASFV, in vitro [27, 29-32].

There is also evidence indicating a role for Natural Killer (NK) cells in protection against ASFV [28]. Thus, pigs that remained asymptomatic after infection with the naturally occurring, non-haemadsorbing and non-fatal ASFV isolate NH/P68 showed increased NK activity 7 days after infection and did not have viraemia or fever after day 14 post-infection. In contrast, NK activity in pigs that developed lesions and had viraemia and fever late after infection, was similar to that of control animals.

Approaches to Vaccine Development

Attenuated vaccines

Immunisation of pigs with naturally attenuated ASFV strains, strains of moderate virulence and strains attenuated by passage in cell culture or by targeted gene deletions have been shown to protect pigs against challenge with virulent viruses [17, 28, 33-36]. To define the molecular basis of attenuation, the complete genome of the Portuguese genotype I naturally attenuated ASFV strain OURT88/3 has been sequenced. This revealed a large deletion of about 8 kbp from a region close to the left genome end compared to genomes of virulent isolates. This encodes 5 copies of multigene family (MGF) 360 and 2 of multigene family 505/530 [37]. Targeted deletion of these genes from Pr4 virulent isolate resulted in increased type I interferon production and stimulation of interferon induced responses [4]. OURT88/3 also has frame shift mutations in the *EP402R* and *EP153R* genes encoding two proteins with roles in immune evasion [38, 39]. These gene deletions and disruptions are likely to be important for attenuation of the virus and may play a role in enhancing the induction of protective immunity.

The growth in information concerning functions of virus-encoded genes and available sequence from many isolates has opened the possibility for rational attenuation of ASFV to produce vaccine strains. Genes whose deletion reduces virus virulence or increases the immune response have been identified by targeted deletion [40]. Deletion of genes which reduces virus replication in macrophages generally reduces virus virulence. These include members of MGF 360 and 530 [4], genes encoding enzymes involved in nucleotide metabolism such as the thymidine kinase (TK) [41] and an ERV 1 homologue (B119L or 9GL) which is part of a redox pathway involved in virus morphogenesis [34]. Deletion of other genes which do not reduce replication in macrophages can also reduce virulence in pigs. For example, deleting the *DP71L* (also named *l14L*, *NL*, *NLS*) gene can reduce virulence depending on the virus isolate as does deletion of the *DP96R* (UK) virulence gene [42, 43]. In contrast, deletion of some genes involved in immune evasion does not reduce virulence. For example, deleting the genes encoding A238L, the CD2v (EP402R), C-type lectin (EP153R) and IAP inhibitor of apoptosis genes did not reduce virulence in pigs [38, 44, 45]. However deletion of at least some of these genes may be advantageous in producing a rationally attenuated vaccine since the expressed proteins may interfere with induction of an appropriate protective response. *A238L* is known to interfere with transcriptional activation of a broad range of host immunomodulatory genes [44, 46, 47]; *CD2v* causes binding of red cells to infected cells and extracellular virions and may inhibit their recognition in blood by components of the immune system [38, 48] and *EP153R* may inhibit presentation of peptides with MHC class I [49].

Studies have shown that immunisation of pigs with attenuated strains can induce protection against related virulent viruses [35, 50]. Immunisation with the attenuated genotype I strain OURT88/3 has been shown to induce protection against challenge with other genotype I isolates and a virulent genotype X isolate from Uganda [31, 33]. In this study, cross-protection was shown to correlate with recognition of different ASFV strains by IFNγ-producing, ASFV-immune lymphocytes. This demonstration of cross-protection between genotypes indicates that candidate vaccine strains could be deployed in regions where more than one virus genotype is circulating, dependent on what those genotypes are.

A disadvantage of using live attenuated vaccines in the field may be the risk of persistence of the vaccine strain. The use of vaccine strains which undergo a single cycle of infection but do not produce infectious progeny would avoid this risk. However a higher virus load would most likely be required to induce protection. This approach also has the advantage that knowledge of protective antigens is not required. Placing genes for essential virus structural proteins, such as that encoding the pp220 polyprotein precursor of core virion proteins under control of an inducible promoter leads to production of virus particles that lack cores and are non-infectious [51, 52]. These would be attractive candidates for production of such a vaccine.

Subunit vaccines

Vaccines based on the delivery of one or more antigens that induce a protective response would have the advantage of avoiding use of a live virus vaccine strain and could use well established platform technologies for vaccine production. However, to achieve this, knowledge of protective antigens is required.

Some progress has been made in identifying proteins which induce a humoral response that can at least induce partial protection. Antibodies to proteins p72 and p54 inhibit virus attachment, while antibodies to protein p30 inhibit virus internalization [53]. Immunization of pigs with a mixture of p54 and p30 proteins, or a chimera of both these proteins, induced virus neutralizing antibodies and drastically modified the disease course, with a variable degree of protection [54, 55]. In contrast a separate study showed that neutralizing antibodies to ASFV proteins p30, p54, and p72 were unable to mediate protection [56]. Recombinant ASFV CD2v protein produced in baculovirus also induced partial protection in pigs [57]. DNA vaccination has also been used as a tool to identify protective antigens. This has the advantage of inducing both cellular and humoral antibody responses. In one approach, immunisation with a plasmid containing a gene fusion of several ASFV genes with a single chain variable fragment of an antibody specific for a swine leukocyte antigen II [58] was shown to amplify the antibody response in pigs.

However the antibodies were non-neutralising and protection was not observed.

Some epitopes recognised by ASFV-specific CTLs have been identified. Target cells infected with a recombinant vaccinia virus (VV) expressing the ASF virus p30, were recognized and lysed by CTL [32]. Expression of ASFV antigens as fusions with *Pseudomonas aeruginosa* outer membrane lipoprotein I gene (oprI) [59] identified MHC-class I restricted epitopes from the major ASFV capsid protein p72 and from ORF G1340L. However a pig immunised with outer membrane preparations from bacteria expressing ORF G1340L was shown not to be protected against challenge with virulent virus [60], even though lymphocytes from the vaccinated animal were able to reduce the yield of ASFV from infected autologous macrophages by >99%, in vitro [60]. Since there appears to be a correlation between protection and ASFV-specific IFNγ-producing T cells [28], further studies are required to identify the ASFV antigens recognised by ASFV-specific memory T cells, which can then be incorporated into a vectored vaccine.

The development of rapid approaches for screening pathogen genomes for protective antigens [61-63] are applicable to ASFV and research in this area is in progress. Using a prime boost strategy [64] such as a DNA prime and boosting with antigen expressed by the Modified Vaccinia Ankara (MVA) strain leads to significantly enhanced immune responses. The development of high throughput methods for construction of vaccinia virus recombinants [65] has enabled construction of a large number of vaccinia virus recombinants expressing individual ASFV antigens and the delivery of pools of these by a DNA prime/recombinant vaccinia virus boost regime is being used to analyse immune responses to individual ASFV antigens and define those which are protective (unpublished results Takamatsu, Jankovich, Chapman, Netherton, Goatley, Taylor, Dixon, Jacobs, Sykes).

Research Gaps

It is likely that both attenuated ASFV vaccines and subunit vaccines, perhaps delivered by virus vectors, would be used in the field in different situations. The former would most likely be used in regions where disease is endemic and the latter in the event of disease introduction to a new region.

Further research is required to better understand the role of individual ASFV genes in virus replication and host interactions. The effect of individual and multiple gene deletions on virus pathogenesis and induction of a protective immune response also needs to be studied. This will support the development of live attenuated vaccines. Knowledge is also required of which antigens can induce a protective response and appropriate vectors for delivery of these to pigs need to be tested. The field use of these vaccines will benefit from the development in parallel of diagnostic tests which can distinguish vaccinated from infected animals (DIVA tests).

Better knowledge is also required of the ASFV strains circulating particularly in Africa and the extent to which vaccines can induce cross-protection between strains.

Acknowledgements

We acknowledge financial support from the BBSRC and DEFRA.

References

1 Dixon LK, Escribano JM, Martins C, Rock DL, Salas ML, Wilkinson PJ: Asfarviridae. In Fauquet CM, Mayo MA, Maniloff J, Desselberge U, Ball LA (eds): Virus Taxonomy, VIIIth Report of the ICTV, Elsevier/Academic Press, London, 2005, pp 135-143.

2 Chapman DAG, Darby AC, Da Silva M, Upton C, Radford AD, Dixon LK: Genomic analysis of highly virulent Georgia 2007/1 isolate of African swine fever virus. Emerg Infect Dise 2011;17(4):599-605.

3 de Villiers EP, Gallardo C, Arias M, da Silva M, Upton C, Martin R, et al: Phylogenomic analysis of 11 complete African swine fever virus genome sequences. Virology 2010;400(1):128-136.

4 Afonso CL, Piccone ME, Zaffuto KM, Neilan J, Kutish GF, Lu Z, et al: African swine fever virus multi-gene family 360 and 530 genes affect host interferon response. J Virol 2004;78(4):1858-1864.

5 Bastos ADS, Penrith ML, Cruciere C, Edrich JL, Hutchings G, Roger F, et al: Genotyping field strains of African swine fever virus by partial p72 gene characterisation. Arch Virol 2003;148(4):693-706.

6 Lubisi BA, Bastos ADS, Dwarka RM, Vosloo W: Molecular epidemiology of African swine fever in East Africa. Arch Virol 2005;150(12):2439-2452.

7 Nix RJ, Gallardo C, Hutchings G, Blanco E, Dixon LK: Molecular epidemiology of African swine fever virus studied by analysis of four variable genome regions. Arch Virol 2006;151(12):2475-2494.

8 Gallardo C, Ademun AR, Nieto R, Nantima N, Arias M, Martin E, et al: Genotyping of African swine fever virus (ASFV) isolates associated with disease outbreaks in Uganda in 2007. African J Biotechnol 2011;10(17):3488-3497.

9 Gallardo C, Anchuelo R, Pelayo V, Poudevigne F, Leon T, Nzoussi J, et al: African swine fever virus p72 genotype IX in domestic pigs, Congo, 2009. Emerg Infect Dis 2011;17(8):1556-1558.

10 Rowlands RJ, Michaud V, Heath L, Hutchings G, Oura C, Vosloo W, et al African swine fever virus isolate, Georgia, 2007. Emerg Infect Dis 2008;14(12):1870-1874.

11 Epifano C, Krijnse-Locker J, Salas ML, Salas J, Rodriguez JM: Generation of filamentous instead of icosahedral particles by repression of African swine fever virus structural protein pB438L. J Virol 2006;80(23):11456-11466.

12 McCullough KC, Basta S, Knotig S, Gerber H, Schaffner R, Kim YB, et al: Intermediate stages in monocyte-macrophage differentiation modulate phenotype and susceptibility to virus infection. Immunology 1999;98(2):203-212.

13 Sanchez-Torres C, Gomez-Puertas P, Gomez-del-Moral M, Alonso F, Escribano JM, Ezquerra A, et al: Expression of porcine CD163 on monocytes/macrophages correlates with permissiveness to African swine fever infection. Arch Virol 2003;148(12):2307-2323.

14 Salas ML, Andrés G: African swine fever virus morphogenesis. Virus Res 2012; Oct 8. doi:pii: S0168-1702(12)00369-3. 10.1016/j.virusres.2012.09.016. [Epub ahead of print]

15 Dixon LK, Alonso C, Escribano JM, Martins C, Revilla Y, Salas ML, Takamatsu H: The Asfarviridae. In: King, AMQ, Adams, MJ, Carstens, EB, Lefkowitz, EJ (eds.), Virus Taxonomy: Classification and Nomenclature of Viruses. Ninth report of the International Committee on Taxonomy of Viruses. Elsevier Aca-demic Press, Amsterdam, pp153–162.

16 Detray DE: Persistence of viremia and immunity in African swine fever. Am J Vet Res 1957;18(69):811-816.

17 Manso-Ribeiro J, Nunes-Petisca JL, Lopez-Frazao F, Sobral M: Vaccination against ASF. Bull OIE 1963;60:921-937.

18 Wardley RC, Norley SG, Martins CV, Lawman MJP: The host response to African swine fever virus. Prog Med Virol 1987;34:180-192.

19 Wardley RC, Norley SG, Wilkinson PJ, Williams S: The role of antibody in protection against African swine fever virus. Vet Immunol Immunopathol 1985;9(3):201-212.

20 Zsak L, Onisk DV, Afonso CL, Rock DL: Virulent African swine fever virus isolates are neutralized by swine immune serum and by monoclonal antibodies recognizing a 72-Kda viral protein. Virology 1993;196(2):596-602.

21 Ruiz Gonzalvo F, Caballero C, Martinez J, Carnero ME: Neutralization of African swine fever virus by sera from African swine fever-resistant pigs. Am J Vet Res 1986;47(8):1858-1862.

22 Borca MV, Irusta P, Carrillo C, Afonso CL, Burrage T, Rock DL: African swine fever virus structural protein P72 contains a conformational neutralizing epitope. Virology 1994;201(2):413-418.

23 Gomez-Puertas P, Oviedo JM, Rodriguez F, Coll J, Escribano JM: Neutralization susceptibility of African swine fever virus is dependent on the phospholipid composition of viral particles. Virology 1997;228(2):180-189.

24 Onisk DV, Borca MV, Kutish G, Kramer E, Irusta P, Rock DL: Passively transferred African swine fever virus-antibodies protect swine against lethal infection. Virology 1994;198(1):350-354.

25 Knudsen RC, Genovesi EV, Whyard TC: In vitro immune serum-mediated protection of pig monocytes against African swine fever virus. Am J Vet Res 1987;48(7):1067-1071.

26 Ruiz Gonzalvo F, Carnero ME, Caballero C, Martinez J: Inhibition of African swine fever infection in the presence of immune sera in vivo and in vitro. Am J Vet Res 1986;47(6):1249-1252.

27 Oura CAL, Denyer MS, Takamatsu H, Parkhouse RME: In vivo depletion of CD8(+) T lymphocytes abrogates protective immunity to African swine fever virus. J Gen Virol 2005;86:2445-2450.

28 Leitao A, Cartaxeiro C, Coelho R, Cruz B, Parkhouse RME, Portugal FC, et al: The non-haemadsorbing African swine fever virus isolate ASFV/NH/P68 provides a model for defining the protective anti- virus immune response. J Gen Virol 2001;82:513-523.

29 Martins CLV, Lawman MJP, Scholl T, Mebus CA, Lunney JK: African swine fever virus specific porcine cytotoxic T-cell Aativity. Arch Virol 1993;129(1-4):211-225.

30 Denyer MS, Wileman TE, Stirling CMA, Zuber B, Takamatsu HH: Perforin expression can define CD8 positive lymphocyte subsets in pigs allowing phenotypic and functional analysis of natural killer, cytotoxic T, natural killer T and MHC un-restricted cytotoxic T-cells. Vet Immunol Immunopathol 2006;110(3-4):279-292.

31 King K, Chapman D, Argilaguet JM, Fishbourne E, Hutet E, Cariolet R, et al: Protection of European domestic pigs from virulent African isolates of African swine fever virus by experimental immunisation. Vaccine 2011;29(28):4593-4600.

32 Alonso F, Dominguez J, Vinuela E, Revilla Y: African swine fever virus-specific cytotoxic T lymphocytes recognize the 32 kDa immediate early protein (vp32). Virus Res 1997;49(2):123-130.

33 Boinas FS, Hutchings GH, Dixon LK, Wilkinson PJ: Characterization of pathogenic and non-pathogenic African swine fever virus isolates from Ornithodoros erraticus inhabiting pig premises in Portugal. J Gen Virol 2004;85:2177-2187.

34 Lewis T, Zsak L, Burrage TG, Lu Z, Kutish GF, Neilan JG, et al: An African swine fever virus ERV1-ALR homologue, 9GL, affects virion maturation and viral growth in macrophages and viral virulence in swine. J Virol 2000;74(3):1275-1285.

35 Hamdy FM, Dardiri AH. Clinical and immunological responses of pigs to African swine fever virus isolated from the Western hemisphere. Am J Vet Res 1984;45(4):711-714.

36 Mebus CA, Dardiri AH: Western hemisphere isolates of African swine fever virus asymptomatic carriers and resistance to challenge inoculation. Am Journal Vet Res 1980;41(11):1867-1869.

37 Chapman DAG, Tcherepanov V, Upton C, Dixon LK: Comparison of the genome sequences of non-pathogenic and pathogenic African swine fever virus isolates. J Gen Virol 2008;89:397-408.

38 Borca MV, Carrillo C, Zsak L, Laegreid WW, Kutish GF, Neilan JG, et al: Deletion of a CD2-like gene, 8-DR, from African swine fever virus affects viral infection in domestic swine. J Virol 1998;72(4):2881-2889.

39 Hurtado C, Jose Bustos M, Granja AG, de Leon P, Sabina P, Lopez-Vinas E, et al: The African swine fever virus lectin EP153R modulates the surface membrane expression of MHC class I antigens. Arch Virol 2011;156(2):219-234.

40 Tulman ER, Delhon GA, Ku BK, Rock DL: African swine fever virus. Curr Top Microbiol Immunol 2009;328:43-87.

41 Moore DM, Zsak L, Neilan JG, Lu Z, Rock DL: The African swine fever virus thymidine kinase gene is required for efficient replication in swine macrophages and for virulence in swine. J Virol 1998;72(12):10310-10315.

42 Zsak L, Caler E, Lu Z, Kutish GF, Neiland JG, Rock DL: A nonessential African swine fever virus gene UK is a significant virulence determinant in domestic swine. J Virol 1998;72(2):1028-1035.

43 Zsak L, Lu Z, Kutish GF, Neilan JG, Rock DL: An African swine fever virus virulence-associated gene NL-S with similarity to the herpes simplex virus ICP34.5 gene. J Virology 1996;70(12):8865-8871.

44 Salguero FJ, Gil S, Revilla Y, Gallardo C, Arias M, Martins C: Cytokine mRNA expression and pathological findings in pigs inoculated with African swine fever virus (E-70) deleted on A238L. Vet Immunol Immunopathol 2008;124(1-2):107-119.

45 Neilan JG, Borca MV, Lu Z, Kutish GF, Kleiboeker SB, Carrillo C, et al: An African swine fever virus ORF with similarity to C-type lectins is non-essential for growth in swine macrophages in vitro and for virus virulence in domestic swine. J Gen Virol 1999;80:2693-2697.

46 Granja AG, Sanchez EG, Sabina P, Fresno M, Revilla Y: African swine fever virus blocks the host cell antiviral inflammatory response through a direct inhibition of PKC-theta-mediated p300 transactivation. J Virol 2009;83(2):969-980.

47 Miskin JE, Abrams CC, Goatley LC, Dixon LK: A viral mechanism for inhibition of the cellular phosphatase calcineurin. Science 1998;281(5376):562-565.

48 Quintero JC, Wesley RD, Whyard TC, Gregg D, Mebus CA: In vitro and in vivo association of African swine fever virus with swine erythrocytes. Am J Vet Res 1986;47(5):1125-1131.

49 Hurtado C, Granja AG, Bustos MJ, Nogal ML, de Buitrago GG, de Yebenes VG, et al: The C-type lectin homologue gene (EP153R) of African swine fever virus inhibits apoptosis both in virus infection and in heterologous expression. Virology 2004;326(1):160-170.

50 Ruiz Gonzalvo F, Carnero ME, Bruyel V: Immunological responses of pigs to partially attenuated African swine fever virus and their resistance to virulent homologous and heterologous viruses. In Wilkinson PJ (ed.): African Swine Fever, Proc CEC/FAO Research Seminar, Sardinia, September 1981, EEC Publication EUR 8466, 1983, pp206–216.

51 Andres G, Garcia-Escudero R, Salas ML, Rodriguez JM: Repression of African swine fever virus polyprotein pp220- encoding gene leads to the assembly of icosahedral core-less particles. J Virol 2002;76(6):2654-2666.

52 Suarez C, Salas ML, Rodriguez JM: African swine fever virus polyprotein pp62 is essential for viral core development. J Virol 2010;84(1):176-187.

53 Gomez-Puertas P, Rodriguez F, Oviedo JM, RamiroIbanez F, RuizGonzalvo F, Alonso C, et al: Neutralizing antibodies to different proteins of African swine fever virus inhibit both virus attachment and internalization. J Virol 1996;70(8):5689-56894.

54 Gomez-Puertas P, Rodriguez F, Oviedo JM, Brun A, Alonso C, Escribano JM: The African swine fever virus proteins p54 and p30 are involved in two distinct steps of virus attachment and both contribute to the antibody-mediated protective immune response. Virology 1998;243(2):461-471.

55 Barderas MG, Rodriguez F, Gomez-Puertas P, Aviles M, Beitia F, Alonso C, et al: Antigenic and immunogenic properties of a chimera of two immunodominant African swine fever virus proteins. Arch Virol 2001;146(9):1681-1691.

56 Neilan JG, Zsak L, Lu Z, Burrage TG, Kutish GF, Rock DL: Neutralizing antibodies to African swine fever virus proteins p30, p54, and p72 are not sufficient for antibody-mediated protection. Virology 2004; 319(2):337-342.

57 Ruiz-Gonzalvo F, Rodriguez F, Escribano JM: Functional and immunological properties of the baculovirus- expressed hemagglutinin of African swine fever virus. Virology 1996;218(1):285-289.

58 Argilaguet JM, Perez-Martin E, Gallardo C, Salguero FJ, Borrego B, Lacasta A, et al: Enhancing DNA immunization by targeting ASFV antigens to SLA-II bearing cells. Vaccine 2011;29(33):5379-5385.

59 Leitao A, Malur A, Cornelis P, Martins CLV: Identification of a 25-aminoacid sequence from the major African swine fever virus structural protein VP72 recognised by porcine cytotoxic T lymphocytes using a lipoprotein based expression system. J Virol Methods 1998;75(1):113-119.

60 Leitao A, Malur A, Cartaxeiro C, Vasco G, Cruz B, Cornelis P, et al: Bacterial lipoprotein based expression vectors as tools for the characterisation of African swine fever virus (ASFV) antigens. Arch Virol 2000;145(8):1639-1657.

61 Stemke-Hale K, Kaltenboeck B, DeGraves FJ, Sykes KF, Huang J, Bu CH, et al: Screening the whole genome of a pathogen in vivo for individual protective antigens. Vaccine 2005;23(23):3016-3025.

62 Sykes K: Progress in the development of genetic immunization. Expert Rev Vaccines 2008;7(9):1395-1404.

63 Ulmer JB, Liu MA: ELI's coming: Expression library immunization and vaccine antigen discovery. Trends Microbiol 1996;4(5):169-170.

64 Schneider J, Gilbert SC, Blanchard TJ, Hanke T, Robson KJ, Hannan CM, et al: Enhanced immunogenicity for CD8+ T cell induction and complete protective efficacy of malaria DNA vaccination by boosting with modified vaccinia virus Ankara. Nature Med 1998;4(4):397-402.

65 White SD, Conwell K, Langland JO, Jacobs BL: Use of a negative selectable marker for rapid selection of recombinant vaccinia virus. Biotechniques 2011;50(5):303-309.

Linda K Dixon, The Pirbright Institute, Ash Road, Pirbright, Woking, Surrey, GU24 0NF, UK
Tel:+44 1483 232441, fax: +44 1483 232448
Email: linda.dixon@iah.ac.uk

Roth JA, Richt JA, Morozov IA (eds): Vaccines and Diagnostics for Transboundary Animal Diseases. Dev Biol (Basel). Basel, Karger, 2013, vol 135, pp 159-165.

African Swine Fever Diagnosis Update

J.M. Sánchez-Vizcaíno, L. Mur

VISAVET Center and Animal Health Department, Veterinary School, Complutense University of Madrid, Madrid, Spain

Key words: African swine fever, diagnosis, livestock diseases

Abstract: African swine fever (ASF) is one of the most complex livestock diseases. The significant losses that it causes, coupled with the lack of a vaccine against ASF virus and the possible resemblance with other swine hemorrhagic diseases, make early detection and laboratory diagnosis essential for controlling and managing the disease.

All the techniques currently used to diagnose ASF have been fully validated showing high sensitivity and specificity to detect both antigen and antibodies against all 22 known genotypes; and enable the correct diagnosis of ASF in all possible epidemiological situations. Because no vaccine is available, the presence of antibodies always indicates previous infection, and serological diagnosis must always be performed in parallel with antigen detection to increase the sensitivity and specificity of the analyses. Recent developments in ASF diagnosis, specifically the new field diagnostic tests, have improved and facilitated the likelihood of ASF early detection, essential to fighting the disease.

Introduction

African swine fever (ASF) is a very complex swine disease caused by a large, double-stranded DNA virus that is the only member of the *Asfarviridae* family. ASF virus (ASFV) is widely distributed on the African continent, where all 22 known ASFV genotypes have been found. The virus is also present in Sardinia (Italy), where genotype I has been endemic since 1978, as well as in some Eastern European countries affected by genotype II since 2007. ASFV can cause severe consequences, with the resulting ASF taking different clinical courses depending on the viral isolate, host, viral dose, and route of entry [1]. The disease can progress from a hyper-acute form, which causes death in 90% of affected animals, to a sub-acute form. Currently, only acute forms of the disease have been found in the Russian Federation and other affected Eastern European countries. On the African continent, however, all possible clinical presentations are observed due to the existence of asymptomatic carrier animals and chronic infections. ASF symptoms and lesions may resemble those of other hemorrhagic diseases such as classical swine fever (CSF), erysipelas, acute salmonellosis and other septicemias. Therefore, laboratory diagnosis is essential for correct differential diagnosis among these hemorrhagic diseases.

Laboratory diagnosis of ASF is well developed: a wide range of techniques has been fully validated for detecting all known ASFV genotypes [2]. In addition, some techniques have recently been developed for rapid diagnosis of ASF in field conditions. Since no vaccine against ASFV is available, the presence of anti-ASFV antibodies always indicates previous infection. These antibodies appear early post-infection and persist for long periods; the viremia also persists due to the lack of neutralizing antibodies. For these reasons, ASF should always be diagnosed through parallel detection of antigen and antibodies in order to fully understand the ASF epidemiological situation in the affected area, determine the approximate date of infection, and identify all infected and carrier animals.

Antigen Detection

The safest, most convenient and most commonly used techniques for ASFV antigen detection are the polymerase chain reaction (PCR), direct immunofluorescence (DIF), and hemadsorption (HA) technique. Since some ASFV isolates are non haemadsorbing, negative HA results should be confirmed by PCR and/or DIF to avoid false negative results. HA is the gold standard technique for antigen detection and must always be performed to confirm the first cases of ASF in ASF-free countries. It is based on a specific characteristic of ASFV that causes erythrocytes to attach around infected macrophages, leading to the appearance of rosettes before the cytopathic effect [3]. The procedure takes a minimum of 7-10 days to complete, and laboratories must adhere to special biosecurity measures and animal facilities; thus, HA is routinely performed only in reference laboratories.

DIF is another widely used test for detecting ASFV antigen in infected tissues. In this test, a conjugated immunoglobulin against ASFV binds to ASFV antigen in smears of infected tissues, specifically spleen, lung, kidney and lymph nodes. It is a rapid, simple, and economical technique that does not require special equipment other than a fluorescence microscope. However, DIF is recommended only for acute and hyper-acute forms of the disease, since its sensitivity is substantially less for sub-acute and chronic forms where anti-ASFV antibodies are present. These antibodies and, specifically, the antigen-antibody complexes formed do not allow the fixation of the conjugate, leading to false negative results. Therefore, DIF should be used in parallel with another serological technique.

Currently, the most widely used technique for ASFV detection is PCR, several protocols for which appear in the Manual of Diagnostic Tests and Vaccines for Terrestrial Animals of the OIE [2]. These protocols serve for conventional PCR [4] including multiplex PCR for differentiating ASF and CSF, and real-time PCR [5]. The wide dissemination of PCR technology, together with its suitability even for poorly preserved samples, has made it a routine approach for laboratories around the world. More PCR protocols have recently been developed to detect ASFV, though they still await full diagnostic evaluation; these include novel real-time PCR methods [6], PCR using Universal Probe libraries [7], loop amplification assays (LAMPs) and linear-after-the-exponential PCR (LATE-PCR). Sequencing the PCR products allows confirmation of the presence of the desired pathogen, increasing the specificity of the technique.

Sequencing, in fact, is traditionally used to differentiate among ASFV isolates. Given that ASF infection does not produce neutralizing antibodies, classifying ASFV isolates requires genotyping instead of serotyping. To date ASFV isolates around the world have been classified into 22 genotypes based on the partial sequence of the VP72 gene (B646L) [8]. These genotypes are distributed heterogeneously around the world.

Genotype I is the most conserved genotype, but historically also the most widely distributed. Currently it is the only ASFV genotype in Sardinia and in Western and Central African countries, and it co-occurs with other genotypes in some Eastern and Southern African countries. It was the first ASFV to escape from Africa, causing outbreaks in Europe, the Caribbean and South America. Then, in 2007, genotype II, which is present in Mozambique, Zambia, Madagascar and Mauritius, arrived on the European continent. It arrived first in Georgia, from which it spread rapidly through Eastern Europe [9].

Studies of the 22 ASFV genotypes in Eastern and Southern African countries have revealed high genetic variability. This seems to be related to the sylvatic cycle present in the region, and it is responsible for the complex epidemiology of the disease in the area [8].

ASFV genotypes have been further differentiated through analyses of sequence variability in genes encoding p53 and p32, as well as in the repeats of the so-called Central Variable Region. These studies have established possible relationships among isolates, detecting the origin of some specific outbreaks. More recently, a genotyping microarray has been developed based on VP72 sequence and the Luminex system [10]; this technique allows ASF diagnosis and ASFV genotyping without sequencing. This method is faster, less expensive and less labor-intensive than conventional genotyping.

All the antigen detection techniques described above work well for all 22 known ASFV genotypes [11].

Antibody Detection

Serological tools are essential in the control and management of ASF. Since no vaccine is available against ASFV, the presence of anti-ASFV antibodies always indicates infection. Furthermore, anti-ASFV antibodies appear early in infection and persist up to several years in infected animals. For these reasons, serological tests are useful for detecting sub-acute chronic forms of disease and carrier animals, making them a superb weapon in control and eradication programs [12].

The enzyme-linked immunosorbent assay (ELISA) that is the prescribed test for international trade, is frequently used to screen large numbers of samples because the technique is easily automated. Several types of ELISA against ASFV have been developed, including indirect ELISA in which wells are coated with soluble ASFV antigen that includes VP72; in-house ELISAs; commercially available, ready-to-use ELISAs [2] and ELISAs that use a recombinant antigen recently developed to detect ASFV in poorly preserved sera [13].

To confirm positive or ambiguous ELISA results, immunoblotting (IB) [14], indirect immunofluorescence (IIF) [15] and immunoperoxidase techniques (IPT) [16] can be used. IB is a highly sensitive technique that is easy to perform and easy to interpret; it allows visualization of antibody binding to different viral proteins. It is recommended for poorly preserved sera [17]. IIF has been used widely to diagnose ASF. In fact, the combination of IFI and DIF allows ASFV detection in 90% of infected animals, making both techniques together an effective method for diagnosing ASF [18], though expertise is needed to interpret the results. IPT is based on the same principle as IFI, except that results are visualized through an enzymatic reaction instead of immunofluorescence. Since IPT is performed in 96-well plates, a large number of samples can be processed simultaneously.

Recent studies carried out within the European ASFRISK project of the 7th Framework Programme (KBBE-2007-1-3-05) demonstrate that current serological techniques work well with all ASFV genotypes currently circulating in the world, including the most variable genotype X and the genotype II currently present in Europe [11].

Sample Collection

Sample collection, storage and transport are critical to ensure accurate laboratory diagnosis. To improve sample collection, research groups have described the use of filter papers to detect ASFV antigen [19] and anti-ASFV antibody in blood. Collecting blood samples with these filter papers simplifies and reduces the costs and risks of sample preservation and transport. Another recent study has described a method for sampling oral fluids to detect anti-ASFV antibody. This approach is easier, safer and potentially more cost-effective than traditional bleeding [20].

New Tools and Future Research

Although ASF laboratory diagnostic methods are diverse and well developed, several new approaches have been developed in recent years in an effort to facilitate early diagnosis in field conditions. This need for early diagnosis is, in fact, a challenge for most animal diseases around the world; and it can be addressed only by training veterinarians and farmers to be vigilant against the possible emergence of disease. A good field test may help detect diseases, but it can never substitute for disease awareness and knowledge.

For antigen detection, simple and rapid techniques have recently been developed to allow diagnosis in the field or in the laboratory, including portable PCR machines (T-COR4TM), LAMP assays and prototype commercial PCR tests. Such rapid detection will allow the implementation of control measures as early as possible, consequently reducing the negative consequences of disease spread.

For antibody detection, pen-side tests based on immunochromatography have shown results comparable with those of conventional serological techniques [21]. These small tools can detect antibodies in serum and blood with the same precision as ELISA, yet they take only 10 minutes and do not require any additional equipment.

Conclusions

Although ASF laboratory diagnostic methods are diverse and well developed, several new approaches have been developed in recent years in an effort to facilitate early diagnosis in field conditions. This need for early diagnosis is, in fact, a challenge for most animal diseases around the world; and it can be addressed only by training veterinarians and farmers to be vigilant against the possible emergence of disease. A good field test may help detect diseases, but it can never substitute for disease awareness and knowledge.

For antigen detection, simple and rapid techniques have recently been developed to allow diagnosis in the field or in the laboratory, including portable PCR machines (T-COR4TM), LAMP assays and prototype commercial PCR tests. Such rapid detection will allow the implementation of control measures as early as possible, consequently reducing the negative consequences of disease spread.

For antibody detection, pen-side tests based on immunochromatography have shown results comparable with those of conventional serological techniques [21]. These small tools can detect antibodies in serum and blood with the same precision as ELISA, yet they take only 10 minutes and do not require any additional equipment.

Acknowledgements

Authors acknowledge EU ASFRIK project (EC, FP7-KBBE-2007-1, Project #211691), the World Organisation for Animal Health and the Community Reference Laboratory for ASF. Lina Mur holds a scholarship from the Spanish Ministry of Education, Culture and Sports (FPU program).

References

1 Sánchez-Vizcaíno JM, Arias M: African swine fever, in: Zimmerman J, Karriker L, Ramirez A, Schwartz K, Stevenson G (ed): Diseases of Swine, 10th edn. Blackwell Publishing Professional, Ames, Iowa, 2012; pp 396-404.

2 OIE: African swine fever, in: Manual of diagnostic tests and vaccines for terrestrial animals 2012. Office International des Epizooties, Paris, France, 2012. Chapter 2.8.1.

3 Malmquist WA, Hay D: Hemadsorption and cytophatic effect produced by African swine fever virus in swine bone marrow and buffy coat culture. Am J Vet Res 1960; 21:104-108.

4 Agüero M, Fernandez J, Sánchez-Mascaraque C, Arias M, Sánchez-Vizcaíno JM: Highly sensitive PCR assay for routine diagnosis of African swine fever virus in clinical samples. J Clin Microbiol 2003; 41:4431-4434.

5 King DP, Reid SM, Hutchings GH, Grierson SS, Wilkinson PJ, Dixon LK, et al: Development of a TaqMan PCR assay with internal amplification control for the detection of African swine fever virus. J Virol Methods 2003;107:53-61.

6 Tignon M, Gallardo C, Iscaro C, Hutet E, Van der Stede Y, Kolbasov D, et al: Development and interlaboratory validation study of an improved new real-time PCR assay with internal control for detection and laboratory diagnosis of African swine fever virus. J Virol Methods 2011;178:161-70.

7 Fernández-Pinero J, Gallardo C, Elizalde M, Robles A, Gómez C, Bishop R, et al: Molecular diagnosis of African swine fever by a new real-time PCR using universal probe library. Transbound Emerg Dis 2012; doi: 10.1111/j.1865-1682.2012.01317.x.

8 Boshoff CI, Bastos ADS, Gerber LJ, Vosloo W: Genetic characterisation of African swine fever viruses from outbreaks in southern Africa (1973–1999). Vet Microbiol 2007;121:45–55.

9 Rowlands RJ, Michaud V, Heath L, Hutchings G, Oura C, Vosloo W, et al: African swine fever virus isolate, Georgia, 2007. Emerg Infect Dis 2008;14:1870-1874.

10 LeBlanc N, Cortey M, Fernandez Pinero J, Gallardo C, Masembe C, Okurut AR, et al: Development of a suspension microarray for the genotyping of African swine fever virus targeting the SNPs in the C-terminal end of the p72 gene region of the genome. Transbound Emerg Dis 2012; doi: 10.1111/j.1865-1682.2012.01359.x.

11 Arias M: Development and validation of sensitive diagnostic techniques, including front-line and pen side tests with the existing field viruses. ASFRISK Symposium. Lisbon, Portugal, 2011.

12 Arias M, Sanchez-Vizcaino JM: African swine fever eradication: The Spanish model. In: Morilla A, Jin K, Zimmerman J (eds), Trends in Emerging Viral Infections of Swine. Iowa State University Press, Iowa, United States of America, 2002; pp 133-139.

13 Gallardo C, Reis AL, Kalema-Zikusoka G, Malta J, Soler A, Blanco E, at al: Recombinant antigen targets for serodiagnosis of African swine fever. Clin. Vaccine Immunol 2009;16:1012-1020.

14. Pastor MJ, Laviada MD, Sanchez-Vizcaino JM, Escribano JM: Detection of African swine fever antibodies by immunoblotting assay. Can J Vet Res 1989;53:105-107.

15 Bool PM, Ordas A, Sanchez-Botija C: Diagnosis of ASFV by immunofluorescence. Bull Off Int Epizoot 1969;72:819-839.

16 Martin R, Soler A, Simón A, Martín E, Arias M, Gallardo C: Validation of serological confirmatory test based on Indirect Immunoperoxidase Test. Workshop on Laboratory Diagnosis of African and Classical Swine Fever (ASF and CSF). Lipica, Eslovenia, 2011.

17 Arias M, Escribano JM, Sanchez-Vizcaino JM: Persistence of African swine fever antibody reactivity on ELISA and immunoblotting assay. Vet Rec 1993;133:89-90.

18 Sanchez Botija C: Peste porcina Africana. Nuevos desarrollos. Rev Sci Tech Off Int Epiz 1982;1:991-1029.

19 Fernandez Pinero J, Gallardo C, Robles A, Gomez C, Soler A, Arias M: Storage and detection of African swine fever virus DNA in clinical material collected on FTA cards. Proceedings of the 21st IPVS Congress. Vancouver, Canada, 2010.

20 Mur L, Gallardo C, Soler A, Zimmermman J, Pelayo V, Sánchez-Vizcaíno JM, Arias M: Potential use of oral fluid samples for African swine fever antibody detection. Proceedings of the 9th International Congress of Veterinary Virology. Madrid, Spain, 2012.

21 Perez T, Gallardo C, Nieto R, De Mia G, Vela C, Bishop R: Development and preliminary validation of a pen-side test based on the use of vp72 protein for ASFV antibody detection. Proceedings of the 6th International Symposium on Emerging and Re-emerging Pig diseases. Barcelona, Spain, 2011.

Prof. José Manuel Sánchez-Vizcaíno, DVN, PhD, Dhc, Catedrático de Sanidad Animal, Director del laboratorio de referencia de la Organización Mundial de Sanidad Animal (OIE), Universidad Complutense, Facultad de Veterinaria, Avenida Puerta del Hierro s/n, 28040 Madrid, Spain.

Tel. +34.91.394.40.82, fax. +34. 91.394.39.08

Email: jmvizcaino@visavet.ucm.es

Roth JA, Richt JA, Morozov IA (eds): Vaccines and Diagnostics for Transboundary Animal Diseases. Dev Biol (Basel). Basel, Karger, 2013, vol 135, pp 167-174.

Classical Swine Fever

V. Moennig[1], P. Becher[1], M. Beer[2]

[1] Institute of Virology, OIE and EU Reference Laboratory for Classical Swine Fever, University of Veterinary Medicine, Hannover, Germany

[2] Institute of Diagnostic Virology, Friedrich-Loeffler-Institut, Greifswald, Germany

Key words: Classical swine fever, laboratory diagnosis, molecular epidemiology, DIVA vaccines

Abstract: Classical swine fever is a serious and economically important transboundary disease threatening pig production globally. The infection may occur in backyard pigs, feral pig populations and domestic pigs. Whereas there are proven control strategies for the latter pig population, control in backyard pigs with poor biosecurity settings or in wild boar populations of high density still poses a problem in some parts of the world. Laboratory diagnostic methods, efficacious vaccines and contingency plans are in place in most industrialised countries. So far modified live vaccines (MLV) are still the first choice for rapid and reliable immune protection. Since antibodies elicited by conventional MLV cannot be distinguished from antibodies after natural infection, considerable efforts are put into the development of a live marker vaccine accompanied by a serological test. Nevertheless, some remaining gaps with respect to the diagnosis of and vaccination against classical swine fever have been identified.

Introduction

Classical swine fever (CSF) is one of the most severe infectious diseases of pigs with high mortality rates. It is caused by a small enveloped RNA virus, the classical swine fever virus (CSFV), belonging to the *Pestivirus* genus of the *Flaviviridae* family. While CSF had been eradicated decades ago from some parts of the world, e.g. Australia and North America, it is still a problem in parts of the Caribbean and South America as well as parts of Africa and Asia. Except for a minor outbreak of

CSF in Lithuania in recent years, the Member States of the European Union (EU) are free from CSF. However, sporadic outbreaks in some Western Balkan countries and the epidemiological situation in countries east of the EU are less favourable. Of particular concern are endemic situations in backyard pig holdings and wild boar populations in some Eastern European countries. For countries free of CSF, there is a constant threat of reintroduction, especially through illegal imports of fresh pork products, tourism, hunting and (illegal) swill feeding. During the last two decades the EU experienced a series of reintroductions of CSF with serious socio-economic consequences. The overall damage amounted to several billion euros. The prime goal of farmers and veterinarians therefore must be to be prepared and to recognize primary outbreaks of CSF as early as possible. A complicating factor is the clinical diagnosis of CSF, which is sometimes hampered by atypical symptoms that can be confused with other porcine infectious diseases, e.g. porcine circovirus 2 infections and/or porcine respiratory and reproductive syndrome. Especially the detection of index cases might be delayed resulting in a prolonged "high risk period", i.e. the time between introduction and detection of CSFV. All the more it is important to have an arsenal of sensitive and specific laboratory diagnostic tools at hand.

Member States of the EU and other CSF-free countries have adopted a non-vaccination policy, i.e. prophylactic vaccination against CSF is banned in principle. In case of severe outbreaks in domestic pigs and in case of CSF epidemics in wild boar, a so-called "emergency vaccination" is possible. However, since emergency vaccination with conventional vaccines is followed by severe trade restrictions, this option so far has only been used for oral vaccination of wild boar, e.g. in Germany and backyard pigs in Romania, respectively. Therefore there is an urgent demand for safe and effective DIVA vaccines (Differentiating Infected from Vaccinated Animals).

Diagnosis of Classical Swine Fever

The importance of a rapid, precise and inexpensive laboratory diagnosis of CSF cannot be overestimated. Apart from an early clinical diagnosis, rapid laboratory diagnosis can help to shorten the high risk period, and thereby prevent spread of CSFV by trade and other contacts. Over the last 20 years techniques for the detection of virus and virus-specific antibodies have improved greatly. Current routine methods are described in detail in the OIE "Manual of Diagnostic Tests and Vaccines for Terrestrial Animals" and in the "EU Diagnostic Manual for Classical Swine Fever (CSF) diagnosis" (2002/106/EC). Both manuals contribute to a rapid, sensitive and standardized diagnosis of CSF. In addition, the EU Reference Laboratory (EURL) and a network of OIE Reference Laboratories enable National Reference Laboratories (NRLs) to acquire a high standard of performance. For almost 25 years, the EURL in Hannover has conducted interlaboratory comparison tests. The number of participating NRLs has steadily increased. Test results are being discussed in annual meetings, where also the latest scientific results with respect to diagnosis, vaccination and epidemiology of CSF are being presented. Forty-nine laboratories from 41 countries participated in last year's test. Results show that EU NRLs and most laboratories from participating Third Countries are well prepared to diagnose CSF. Laboratories that wished to improve their results performed corrective actions followed by a second round of testing. Ideally NRLs train their regional laboratories to ensure a high quality of CSF diagnosis throughout the country. In addition, several NRLs are performing laboratory exercises in order to assess and verify their preparedness in case of disease emergency.

Detection of virus

The classical method for the detection of CSFV is virus isolation using susceptible porcine cells, which are being co-cultivated with buffy coat cells or sera of infected pigs or organ suspensions of dead animals. Virus isolation is sensitive, highly specific, however, time-consuming and labour intensive. Safety issues related to the shipment of infective samples are additional problems associated with virus isolation. Therefore it was recently proposed by Becher and Postel (personal communication) to extract viral RNA for shipment. RNA samples are noninfectious for animals and they can be used for in vitro electroporation of susceptible cells in order to isolate the original virus. In addition these samples are suitable for nucleic acid amplification techniques. It is also possible to store the samples by inactivating and conserving lysis buffers. RNA isolation can subsequently be done in the receiving laboratory.

For the rapid detection of viral antigen, immunofluorescence and antigen capture enzyme-linked immunosorbent assays (ELISAs) are being used. The former is an established routine method that requires thin sections of autopsy material, e.g lymph nodes, tonsils, spleen, and other organs. Depending on the reagents used, its specificity is good, however, the sensitivity of the assay is not very high. Antigen detection in live animals is possible using the antigen-capture ELISA. Most commercially available assays are in the classical microtitre plate format. The accuracy of test results largely depends on the stage of infection when blood samples were taken. Samples taken during early or late stages may lead to false negative results, since the virus titre in the blood is too low to be detected by these assays. However, some commercial antigen-capture-ELISAs are nearly as sensitive as virus isolation (e.g. panpesti ERNS-capture ELISA) and both methods also still have their place in the diagnosis of CSF because positive results are quickly obtained. When there is a clinical suspicion, negative results should be reinvestigated using a more sensitive method.

Nucleic acid amplification techniques have been adapted for the diagnosis of CSF and polymerase chain reaction after reverse transcription (RT-PCR) is considered the most sensitive and highly specific tool for the detection of CSFV and other pestiviruses. Analysis of amplicons is either performed by gel electrophoresis or a magnetic bead microarray technique, which allows the rapid detection and identification of different pestiviruses. The latter process utilizes an oligonucleotide array, to which 5′ biotinylated PCR products are hybridized, followed by visualization with streptavidin-coated magnetic particles by microscope, biochip reader or even by the naked eye. The relative simplicity of the test makes it suitable for use even in modestly equipped laboratories [1]. The most widespread method, however, is the real time RT-PCR. There are numerous in-house protocols and commercial kits which proved to be highly sensitive and specific [2]. In situations where modified live vaccines (C-strain) are being used to control CSF, a specific real-time RT-PCR discriminating vaccine and field virus is available [3, 4]. It should be feasible to adapt these tests to portable devices.

Loop-mediated isothermal amplification after reverse transcription (RT-LAMP) assays require only modestly equipped laboratories. However, they apparently failed to reach the popularity of real time RT-PCR for the diagnosis of CSFV. A possible reason might be the variability of RNA viruses that could impair the efficiency of the primers.

The EURL in Hannover maintains a large virus collection and database of CSFV sequences that can be used for the molecular epidemiology of CSF by phylogenetic analysis. During the last two decades this database has been used extensively and the number of records has increased significantly. It has proven to be a useful tool for tracing CSFV and the implementation of control measures [5]. However, the resolution of the phylogenetic analyses was limited, since only short sequences were compared, e.g., fragments of the 5' NTR (150 nucleotides, nt) and the E2 gene (190 nt). This made the differentiation of closely related isolates impossible. Recently it was shown that full-length E2 encoding sequences (1119 nt) are more suitable for reliable and statistically significant phylogenetic analysis. The high resolution of this method was demonstrated by the phylogenetic analysis of closely related CSFV isolates from recent CSF outbreaks in Lithuania, Cuba and Nepal [6-8]. Furthermore, next generation sequencing technologies allow the analysis of full-length genomes of CSFV, providing the best basis for a high resolution phylogenetic analysis as it was shown recently by Leifer and co-workers [9]. The EURL therefore recommends to apply these strategies as they provide an improved basis for CSF viral molecular epidemiology.

Detection of antibodies

CSFV-infected pigs mount a strong humoral immune response against the virus and first antibodies are detectable at about 10-14 days post-infection. The investigation of virus-specific antibodies is very important in CSF control and surveillance schemes. In addition, the distinction of CSFV-specific antibodies from antibodies directed against ruminant pestiviruses and the differentiation of vaccinated and field-virus infected animals are of great importance. The gold standard for CSF serology is the virus neutralization assay (VNA). The VNA is sensitive, specific and – to a certain extent - it allows for the differentiation of pestiviruses. However, the tests are performed in cell culture, infected cells must be visualized by immunostaining and therefore these procedures are time-consuming and labour-intensive. Another disadvantage of the VNA is its inability to distinguish antibodies elicited by DIVA vaccines from antibodies against field virus. In contrast, ELISAs in microtitre formats are easy to perform, rapid and automated. In general they are not as sensitive and specific as the VNA, but a distinction between antibodies against different pestiviruses is possible and depends on the design of the test [10]. Likewise depending on the test design there are ELISAs which are able to differentiate immune responses against natural CSFV infections from antibodies induced by DIVA vaccines. However, sensitivity and specificity of these discriminating tests are not satisfactory in all cases and discriminatory assays should be subject of continuous improvement.

A further development of the classical antibody ELISA in the microtitre format is the immunochromatographic strip or lateral flow device which requires no specific laboratory environment and can be used as a "penside test". These tests are using a solid phase membrane with test and control lines coupled with absorbent pads. The test strips can either be dipped into the sample (e.g. blood, serum or meat juice) or the sample may be applied to a designated area of the strip [11].

Vaccination against Classical Swine Fever

Pigs surviving natural CSFV infection are immune for the rest of their lives. Similarly, modified live vaccines (MLV), e.g. the lapinised Chinese (C-) strain of CSFV elicit high titres of protective antibodies which ensure a sterile and long lasting immunity. Most MLV against CSF are highly efficacious [12], safe and innocuous [13]. They have been used during the last decades to control CSF in endemic situations and systematic vaccination often was the first step towards CSF eradication. CSFV vaccines and their safety when used in emergencies have been reviewed extensively by the Scientific Panel of the European Food Safety Authority [14]. Oral immunization of wild boar using baits with MLV has been very successful to control and eradicate CSF from wild boar populations. Unfortunately, the antibodies induced by CSFV MLV cannot be distinguished from those induced by natural infection. As a consequence, products of vaccinated animals are excluded from international trade and the vaccinated area is considered not to be free from CSF. In contrast, derogations from the restrictions are possible in case DIVA vaccines have been used (Council Directive 2001/89/EC). In order to overcome these restrictions, a first generation of baculovirus-expressed E2 glycoprotein marker vaccines had been developed [15, 16]. These vaccines elicited protective antibodies, but required two immunisations and immunity could only be expected about three weeks after primary vaccination. Using these vaccines did not lead to a complete block of transplacental infection as observed after MLV application. Furthermore these vaccines are unsuitable for oral immunization of wild boar or backyard pigs. In order to achieve earlier protection and make oral vaccination possible, a number of attempts have been made to develop live DIVA vaccines. The experimental approaches focused on DNA vaccines, viral vectors expressing CSFV proteins, chimeric pestiviruses, and trans-complemented deleted CSFV genomes (replicons) [17]. Chimeric pestiviruses are promising candidates for a new generation of CSF DIVA vaccines and the goal of an EU-funded research project is the development and licensing of the chimeric pestiviral marker vaccine candidate CP7_E2alf (http://www.csfvaccine.org/). The parental virus is the cytopathic bovine viral diarrhea virus CP7 expressing the E2 protein of CSFV strain Alfort 187. The vaccine candidate virus fulfills the requirements of the World Organisation for Animal Health (OIE) with respect to protection and duration of immunity when applied intramuscularly. The vaccine can also be administered orally, however, as seen for other MLV CSFV vaccines, a few animals do not respond [18]. Serological discrimination of CP7_E2alf-vaccinated animals is possible by detection of antibodies against the CSFV Erns envelope protein [19], because the vaccine does not elicit antibodies against CSFV Erns. In case of natural CSFV infection or superinfection of a vaccinated animal, the discriminatory ELISA would turn positive [10].

It may be expected that CP7_E2alf will be the first live DIVA vaccine for CSF submitted for licensing.

Gaps and Future Research

Recently, a gap analysis has been performed by the EURL in cooperation with other European NRLs and was published on the DISCONTOOLS website (http://www.discontools.eu). The most important issues with respect to diagnosis and vaccination are:

Diagnosis

- There are no reliable and easy-to perform penside tests. Further development and assessment of affordable penside tests for virus and antibodies is necessary. A consensus on the use of penside tests for the detection of CSFV is needed.
- The potential of full length in-depth sequencing for a higher resolution in molecular epidemiology should be explored in more detail.
- The methods for serological differentiation of CSFV infection from other pestivirus infections need to be refined.
- Sensitivity and specificity of serological DIVA ELISAs should be improved.
- The market potential for CSFV diagnostics is small, therefore availability of certain test kits is sometimes critical. Likewise there are no commercial incentives to develop new tests or improve existing ones.

Vaccination

- The immune response against "C-strain" viruses has been assessed for only a few commercial vaccines. Little is known about efficacy and safety of many locally available MLVs vaccines. Further studies on the quality of vaccines which are in use worldwide is necessary.

- The understanding of mechanism of early protection after vaccination with MLV (e.g. C-strain) and live DIVA vaccines should be improved.
- The development and production of "new" type of baits for young wild boar is not yet completely solved.
- Improvement of strategies to eradicate CSF in backyard holdings should be devised including oral vaccination.
- An accurate assessment of the field efficacy of the currently available DIVA vaccine under CSF outbreak conditions has not yet been performed.
- Efficacious and safe live DIVA vaccines suitable for vaccination of domestic pigs and oral administration for wild boar and backyard pigs should be developed and licensed. The license should only be valid in conjunction with the availability of a discriminating DIVA ELISA.
- Since future live DIVA vaccines will most likely be genetically modified organisms, there are uncertainties concerning their registration. The use of genetically modified vaccines might be problematic in some countries.
- Commercial incentives for development and licensing of live DIVA vaccines are lacking, therefore funding of research activities by non-commercial organisations is absolutely essential.

Acknowledgements

Part of the research referred to in this paper was supported by grant KBBE-227003 (CSFV_goDIVA) of DG Research of the European Commission. EURL is supported by DG Health and Consumers (SANCO).

References

1 Leblanc N, Gantelius J, Schwenk JM, Stahl K, Blomberg J, Andersson-Svahn H, et al: Development of a magnetic bead microarray for simultaneous and simple detection of four pestiviruses. J Virol Methods 2009;155(1):1-9.

2 Hoffmann B, Blome S, Bonilauri P, Fernandez-Pinero J, Greiser-Wilke I, Haegeman A, et al: Classical swine fever virus detection: results of a real-time reverse transcription polymerase chain reaction ring trial conducted in the framework of the European network of excellence for epizootic disease diagnosis and control. J Vet Diagn Invest 2011;23(5):999-1004.

3 Liu L, Xia H, Everett H, Sosan O, Crooke H, Meindl-Bohmer A, et al: A generic real-time TaqMan assay for specific detection of lapinized Chinese vaccines against classical swine fever. J Virol Methods 2011;175(2):170-174.

4 Blome S, Gabriel C, Staubach C, Leifer I, Strebelow G, Beer M: Genetic differentiation of infected from vaccinated animals after implementation of an emergency vaccination strategy against classical swine fever in wild boar. Vet Microbiol 2011;15153(3-4):373-376.

5 Dreier S, Zimmermann B, Moennig V, Greiser-Wilke I: A sequence database allowing automated genotyping of classical swine fever virus isolates. J Virol Methods 2007;140(1-2):95-99.

6 Postel A, Schmeiser S, Bernau J, Meindl-Boehmer A, Pridotkas G, Dirbakova Z, et al: Improved strategy for phylogenetic analysis of classical swine fever virus based on full-length E2 encoding sequences. Vet Res 2012;43(1):50.

7 Postel A, Jha VC, Schmeiser S, Becher P: First molecular identification and characterization of classical swine fever virus isolates from Nepal. Arch Virol 2012 (in press).

8 Postel A, Schmeiser S, Perera CL, Rodríguez LJP, Frias-Lepoureau MT, Becher P: Classical swine fever virus isolates from Cuba form a new subgenotype 1.4. Vet Microbiol 2013;161(3-4):334-338.

9 Leifer I, Hoeper D, Blome S, Beer M, Ruggli N: Clustering of classical swine fever virus isolates by codon pair bias. BMC Res Notes 2011;4:521.

10 Aebischer A, Müller M, Hofmann MA: Two newly developed Erns-based ELISAs allow the differentiation of Classical Swine Fever virus-infected from marker-vaccinated animals and the discrimination of pestivirus antibodies. Vet Microbiol 2013;161(3-4):274-285.

11 Li X, Wang L, Shi X, Zhao D, Yang J, Yang S, et al: Development of an immunochromatographic strip for rapid detection of antibodies against classical swine fever virus. J Virol Methods 2012;180(1-2):32-37.

12 Graham SP, Everett HE, Haines FJ, Johns HL, Sosan OA, Salguero FJ, et al: Challenge of pigs with classical swine fever viruses after C-strain vaccination reveals remarkably rapid protection and insights into early immunity. PLoS One 2012;7(1):e29310.

13 Kaden V, Lange E, Kuster H, Muller T, Lange B: An update on safety studies on the attenuated "RIEMSER Schweinepestoralvakzine" for vaccination of wild boar against classical swine fever. Vet Microbiol 2010;143(2-4):133-138.

14 EFSA Panel: Animal health safety of fresh meat derived from pigs vaccinated against Classic Swine Fever. EFSA J 2009;doi:10:2903/j.efsa.2009.933.

15 Bouma A, de Smit AJ, de Kluijver EP, Terpstra C, Moormann RJ: Efficacy and stability of a subunit vaccine based on glycoprotein E2 of classical swine fever virus. Vet Microbiol 1999;66(2):101-114.

16 Konig M, Lengsfeld T, Pauly T, Stark R, Thiel HJ: Classical swine fever virus: independent induction of protective immunity by two structural glycoproteins. J Virol 1995;69(10):6479-6486.

17 Beer M, Reimann I, Hoffmann B, Depner K: Novel marker vaccines against classical swine fever. Vaccine 2007;25(30):5665-5670.

18 Gabriel C, Blome S, Urniza A, Juanola S, Koenen F, Beer M: Towards licensing of CP7_E2alf as marker vaccine against classical swine fever - Duration of immunity. Vaccine. 2012;30(19):2928-2936.

19 Meyer D, Aebischer A, Müller M, Grummer B, Greiser-Wilke I, Moennig V, et al: New insights into the antigenic structure of the glycoprotein Erns of classical swine fever virus by epitope mapping. Virology 2012;in press.

Prof. Dr. Volker Moennig, Institute of Virology, Department of Infectious Diseases, University of Veterinary Medicine Hannover, Bünteweg 17, D-30559 Hannover, Germany

Email: volker.moennig@tiho-hannover.de

Roth JA, Richt JA, Morozov IA (eds): Vaccines and Diagnostics for Transboundary Animal Diseases. Dev Biol (Basel). Basel, Karger, 2013, vol 135, pp 175-182.

Schmallenberg Virus

K. Wernike, B. Hoffmann, M. Beer

Institute of Diagnostic Virology, Friedrich-Loeffler-Institut (FLI), Greifswald–Insel Riems, Germany

Key words: Schmallenberg virus, Orthobunyavirus, congenital malformations, cattle, sheep, goat

Abstract: In late 2011 a novel virus of the Simbu serogroup, family *Bunyaviridae*, genus *Orthobunyavirus*, was discovered at the German-Dutch border and named "Schmallenberg virus" (SBV). Since then, the virus has spread rapidly to other European countries, and the involvement of biting midges has been demonstrated. Affected adult ruminants show none or only mild clinical signs including fever, reduced milk production or diarrhoea. However, in keeping with related viruses, an infection of SBV-naive cows and ewes during a critical period of pregnancy can lead to severe foetal malformations summarized as "arthrogryposis-hydranencephaly syndrome." Diagnostic assays like real-time RT-PCR and antibody ELISA are now available; and substantial knowledge of this virus has been gathered quickly. However, further research and exchange of information are essential.

Introduction

In summer and fall 2011, a hitherto unidentified disease was reported in dairy cattle in North Rhine-Westphalia, Germany and the Netherlands. Clinical signs included fever, decreased milk production and diarrhoea for a few days. Some cattle diseases previously relevant in Europe, such as bluetongue, bovine viral diarrhoea, bovine herpes virus type 1 infection and foot-and-mouth disease, as well as exotic diseases (epizootic haemorrhagic disease, Rift Valley fever and bovine ephemeral fever virus) were excluded.

In November 2011 the causative agent, Schmallenberg virus (SBV), was discovered by a metagenomic approach in samples from acutely diseased cows from a farm in Schmallenberg; in addition, the virus was isolated from the blood of a diseased cow and named after the village from which the first samples were derived [1]. First discovered in adult cows with a mild transient disease, SBV has since been implicated in many cases of severely malformed bovine and ovine offspring [2-4]. Starting from Germany and the Netherlands, the virus spread rapidly to several European countries [5]. In 2012, acute cases in adult cattle were reported from regions which had remained free during the previous season, such as Switzerland, parts of France and the United Kingdom [6-9]. Only recently, antibodies to SBV were detected in serum samples of goats in western Poland, as well as in cattle and sheep in Austria [10, 11].

Taxonomy and Molecular Characteristics

The earliest phylogenetic analyses suggested that SBV is a member of the Simbu serogroup within the genus *Orthobunyavirus*, family *Bunyaviridae* [1]. The *Orthobunyavirus* genus contains over 170 viruses of different veterinary and medical relevance [12]. The Simbu serogroup also includes Akabane virus (AKAV) and Aino virus [1,13]. Further sequence analyses of SBV and other orthobunyaviruses showed that SBV belongs, together with the Australian Douglas virus, to the species Sathuperi virus within the Simbu serogroup [13].

The tripartite RNA genome of orthobunyaviruses comprises a small (S), a medium (M) and a large (L) segment. The S segment encodes the viral nucleocapsid protein (N) and a non-structural protein (NSs); the M segment encodes the precursor polyprotein which is cleaved into the two viral envelope glycoproteins (Gn and Gc) in addition to a non-structural protein (NSm). The L segment encodes the viral polymerase [14, 15]. As with other segmented viruses, reassortment may occur between different orthobunyaviruses [15]; however, studies of the Simbu serogroup are limited. SBV was suggested to be one of the ancestors of Shamonda virus (SHAV), while SHAV is a reassortant of SBV and an unclassified virus [13].

Pathogenesis, Clinical Signs and Immunology

All Simbu serogroup viruses closely related to SBV are arthropod-borne viruses transmitted by mosquitoes and/or *Culicoides* biting midges that mainly cause diseases in ruminants [16-18]. The same transmission mode is considered for SBV as well. Based on similarities between bluetongue virus and SBV regarding affected host species and the pattern of disease spread, *Culicoides* midges were suspected to be at least partially responsible for the spread of SBV [19]. The SBV genome has recently been detected in field collected *C. obsoletus* s.s. in Denmark [20] and *C. obsoletus* s.s. and *C. dewulfi* in Belgium [21, 22].

So far, knowledge about the effects of SBV infection on the mammalian hosts is sparse. However, it can be assumed that many similarities with other related Simbu serogroup viruses exist. SBV infections have been reported in cattle, sheep, goats, alpaca and a bison [22, 23]. In adult ruminants clinical signs are either absent or non-specific with duration of barely one week. In cattle fever, reduction in milk yield (up to 50%) and in some cases diarrhoea were observed [24]. Based on a preliminary experimental infection study with three calves, the viraemia in cattle seems to be short (Fig. 1). Three animals were inoculated with SBV and all of them had positive PCR results between days 2 and 5 after infection [1]. The duration of viraemia after SBV infections is concordant with the viraemic period observed for related viruses. The viraemia of AKAV in naturally infected cattle is short-lived as well, and viral RNA is detectable for generally three to four days, with a maximum of nine days [25, 26]. Although infected adult ruminants showed only mild and transient symptoms, an infection with AKAV during a critical period of pregnancy could have serious effects on the foetus. Besides miscarriage and the birth of mummified foetuses, premature or still birth and the birth of malformed lambs and calves are observed [27, 28]. If the dam is infected during the first month of gestation, AKAV is usually not transferred from the dam to the embryo. Later on, however, the virus may cause varying degrees of foetal malformation mainly in the brain, spinal cord and skeletal muscle. The virus replication in maturing neurons and muscle cells may lead to cell death and inflammation in the surrounding tissue resulting in the observed abnormalities [29]. After infection of pregnant animals with Aino virus, abortions and malformed calves occurred as well [30, 31]. Foetal malformations induced by viruses of the Simbu serogroup are described as arthrogryposis-hydranencephaly syndrome (AHS) [32]. Severity of AHS varies dependent on the gestation period; an early infection may lead to por- or hydran-encephaly, thereafter to torticollis and arthrogryposis. In cattle the critical phase is approximately between days 75 and 175 of gestation, in sheep the period for the occurrence of foetal abnormalities has been shown to vary from day 30 to day 50 [33-36]. Knowledge about AKAV infections in pregnant goats is scarce, but abnormalities have been described [37, 38].

As suggested by the close relationship of SBV to viruses of the Simbu serogroup, typical malformations were reported from affected regions in Europe. SBV has been detected in stillborn or new-born lambs, calves and goat kids [22, 39, 40]; however, the vulnerable gestation period has not been determined yet. The most common malformations comprised arthrogryposis, torticollis, scoliosis, kyphosis, brachygnathia (Fig. 1), poliomyelitis, mild to severe hypoplasia of the central nervous system and cerebellar and narrow spinal cords [4, 41-45]. Lung hypoplasia was also observed [24]. Most affected lambs were born at term, several were stillborn but some deformed lambs were born alive. Even dams that gave birth to one malformed and one healthy lamb/kid within the same litter were reported [40, 42]. These clinical findings, however, were not the manifestation of an acute disease; malformation in foetuses and newborns is seen several months after acute infection of the dam (Fig. 1). Due to the deformations of the lambs, dystocia during parturition occurred frequently; some ewes even died as a consequence of uterine perforation [24, 42].

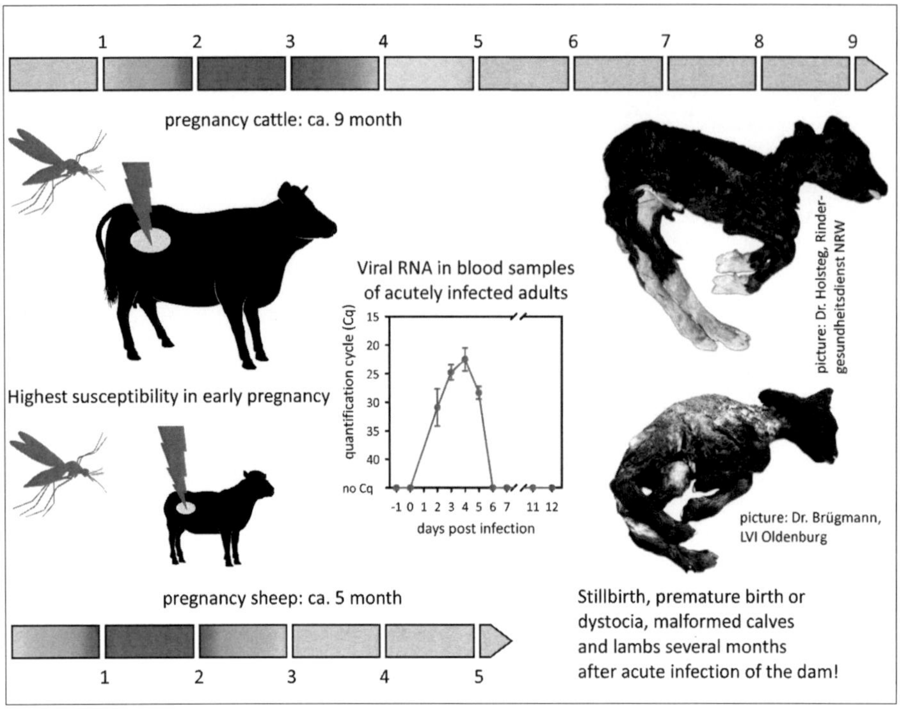

Fig. 1: Genesis of malformations in calves and lambs after infection with Schmallenberg virus during a critical period of pregnancy.

After infection SBV-naive dams develop a mild transient disease associated with viremia for a few days. Dependent on the stage of pregnancy, foetal infections can led to stillbirth, premature birth, dystocia or varying degrees of malformation [63, modified].

Owing to its recent discovery, knowledge about duration of immunity against SBV is scarce. However, the protective neutralizing antibody response in cattle against the closely related AKAV persists for at least two years [46]. Hence, the high within-herd seroprevalence of antibodies against SBV of up to 99% in cows and ewes exposed to infection in 2011 [47-49] suggests an efficacious natural immunity.

So far, zoonotic potential has not been reported from either closely related viruses or from SBV. No SBV-specific antibodies have been detected in any human serum samples collected from persons working or living on SBV-infected farms or from veterinarians [50, 51].

Diagnosis

In most cases knowledge of emerging pathogens is sparse; the preferred sample material is unknown; and specific validated diagnostic tools and protocols are lacking. For SBV-specific genome detection, real-time reverse transcription polymerase chain reaction (RT-qPCR) assays have been developed rapidly [1, 52]. Furthermore, the virus can be isolated on insect and hamster cell lines. For the detection of antibodies, indirect immunofluorescence tests, microneutralization tests and in particular commercial SBV antibody ELISA have all been used (47, 53). As viraemia in acutely infected cattle is short-lived [1], antibody detection is more promising than antigen detection, e.g. for evaluating SBV-prevalence. The preferred sample material for detecting acute infections in adult ruminants and for antibody detection is serum, which also allows for investigation in a neutralisation test. For ELISA or indirect immunofluorescence tests, EDTA plasma samples can also be used.

Suitable sample material of malformed lambs and calves was deduced from other Simbu serogroup viruses. Since the closely related AKAV was isolated from different parts of the brain of affected new-born calves [54], various brain samples were tested by RT-qPCR. It was demonstrated that a high concentration of SBV-RNA can be detected in the cerebrum and the brainstem [52, 55]. Apart from the brain, samples of spleen, meconium, spinal cord, cartilage, umbilical cord or amniotic fluid were tested among others. Amniotic fluid gave positive results in a high percentage of the malformed new-borns, which is of special importance as it can be easily collected by hair swabs without necessity for a necropsy [52].

Consequently, the World Organisation for Animal Health (OIE) has recognized that in addition to organs, blood, placenta and meconium, amniotic fluid from malformed foetuses and new-borns is a suitable sample for viral genome detection [56].

However, not all lambs and calves showing typical malformations had tested positive by RT-qPCR, whereas the majority of the mothers were antibody positive and SBV-infected during pregnancy [57-59]. A possible explanation is the gradual elimination of the virus once the foetus is capable of producing specific neutralizing antibodies in the second part of the gestation period, which results in negative PCR tests [60]. After experimental infection with AKAV, immune-competence is attained after approximately 65–75 days of gestation in lambs and 150 days in calves [61]. As the ruminant placenta prevents the transfer of maternal antibodies during pregnancy, pre-colostral antibodies that may be detected will have been produced by the foetus itself due to infection in utero [62]. Consequently, pre-colostral sera may be a valuable tool to diagnose foetal SBV-infection as well. Indeed, van Maanen and colleagues [57] demonstrated a high percentage (74%) of foetal heart blood samples positive for SBV antibodies, compared to only 27% tested positive by RT-qPCR. Comparable results were obtained using the neutralization test and a commercial ELISA [57].

In conclusion, viral genome detection from different brain samples, placenta, meconium and hair swabs, in combination with antibody detection in heart blood or pre-colostral serum samples seems to be particularly promising to diagnose SBV as the causative agent of malformations in ruminants.

Concluding Remarks

Less than one year ago, a previously unknown arthropod-born disease emerged in Europe and spread very rapidly. Since then, substantial knowledge has been acquired about SBV; initial diagnostic tools have been quickly developed; and data and reagents have been rapidly released and shared globally. In addition, detection by a metagenomics approach proved to be highly efficient; and the veterinary science community now appears to be well prepared for the development of molecular diagnostics for a novel virus.

Nevertheless, many questions remain unanswered. The origin of SBV, possible ways of introduction, the exact distribution in Europe and the responsible vectors are all still unknown or not completely understood. Furthermore, early prototype vaccines have been developed and tested, but no licensed products have yet become available. Finally, the pathogenesis of the Simbu serogroup viruses, especially for the foetus, is not well understood; and this is an important field for future research.

Therefore, intensive research and development, and a continuous exchange of information are of significant importance.

References

1 Hoffmann B, Scheuch M, Hoper D, Jungblut R, Holsteg M, Schirrmeier H, et al: Novel orthobunyavirus in cattle, Europe, 2011. Emerg Infect Dis 2012;18(3):469-472.

2 Dominguez M, Hendrikx P, Zientara S, Calavas D, Jay M, Touratier A, et al: Preliminary estimate of Schmallenberg virus infection impact in sheep flocks - France. Vet Rec 2012; 171(7):426.

3 Herder V, Wohlsein P, Peters M, Hansmann F, Baumgartner W: Salient Lesions in domestic ruminants infected with the emerging so-called Schmallenberg virus in Germany. Vet Pathol 2012; 49(4):588-591.

4 Garigliany MM, Hoffmann B, Dive M, Sartelet A, Bayrou C, Cassart D, et al: Schmallenberg virus in calf born at term with porencephaly, Belgium. Emerg Infect Dis 2012;18(6):1005-1006.

5 Anonymous: European Food Safety Authority publishes its second report on the Schmallenberg virus. Eurosurveill 2012;17(4):pii=20140.

6 Anonymous: Acute cases of SBV found in cattle in southern England. Vet Rec 2012;171(13):307.

7 ProMED-mail: Schmallenberg virus - Europe (48): Switzerland, 1st case, Archive Number: 20120721.1210343. http://www.promedmail.org [posted 22 July 2012].

8 ProMED-mail: Schmallenberg virus - Europe (57): Switzerland, France, Archive Number: 20120907.1285365. http://www.promedmail.org [posted 07 September 2012].

9 Anonymous: Schmallenberg virus 'still circulating' in the UK. Vet Rec 2012;171(6):140.

10 ProMED-mail: Schmallenberg virus - Europe (60): Austria, UK, Belgium, Archive Number: 20120926.1310236. http://www.promedmail.org [posted 26 September 2012].

11 ProMED-mail: Schmallenberg virus - Europe (61): Poland, Archive Number: 20121001.1318740. http://www.promedmail.org [posted 01 October 2012].

12 Plyusnin A, Beaty BJ, Elliott RM, Goldbach R, Kormelink R, Lundkvist A, et al: Bunyaviridae. In: King AMQ, Adams MJ, Carstens EB, Lefkowits EJ (eds) Virustaxonomy: ninth report of the International Committee on Taxonomyof Viruses Elsevier Academic Press, London. 2012; pp.725–741.

13 Goller KV, Höper D, Schirrmeier H, Mettenleiter T, Beer M: Schmallenberg virus as possible ancestor of Shamonda virus. Emerg Infect Dis [Internet] http://dxdoiorg/103201/eid1810120835. 2012.

14 Walter CT, Barr JN: Recent advances in the molecular and cellular biology of bunyaviruses. J Gen Virol. 2011;92(Pt 11):2467-2484.

15 Kobayashi T, Yanase T, Yamakawa M, Kato T, Yoshida K, Tsuda T: Genetic diversity and reassortments among Akabane virus field isolates. Virus Res 2007;130(1-2):162-171.

16 Yanase T, Maeda K, Kato T, Nyuta S, Kamata H, Yamakawa M, et al: The resurgence of Shamonda virus, an African Simbu group virus of the genus Orthobunyavirus, in Japan. Arch Virol. 2005;150(2):361-369.

17 Yanase T, Kato T, Aizawa M, Shuto Y, Shirafuji H, Yamakawa M, et al: Genetic reassortment between Sathuperi and Shamonda viruses of the genus Orthobunyavirus in nature: implications for their genetic relationship to Schmallenberg virus. Arch Virol 2012;157(8):1611-1616.

18 Saeed MF, Li L, Wang H, Weaver SC, Barrett AD: Phylogeny of the Simbu serogroup of the genus Bunyavirus. J Gen Virol 2001;82(Pt 9):2173-2181.

19 Anonymous: Scenarios for the future spread of Schmallenberg virus. Vet Rec 2012;170(10):245-246.

20 Rasmussen LD, Kristensen B, Kirkeby C, Rasmussen TB, Belsham GJ, Bodker R, et al: Culicoids as vectors of schmallenberg virus. Emerg Infect Dis 2012;18(7):1204-1206.

21 ProMED-mail: Schmallenberg virus – Europe (26): vector, morphology, Archive Number: 20120311.1066949. http://www.promedmail.org [posted 11 March 2012].

22 European Food Safety Authority: "Schmallenberg" virus: Analysis of the epidemiological data and assessment of impact. EFSA Journal 2012;10(6):2768 [89 pp] 2012.

23 Jack C, Anstaett O, Adams J, Noad R, Brownlie J, Mertens P: Evidence of seroconversion to SBV in camelids. Vet Rec 2012;170(23):603.

24 Lievaart-Peterson K, Luttikholt SJM, Van den Brom R, Vellema P: Schmallenberg virus infection in small ruminants - First review of the situation and prospects in Northern Europe. Small Ruminant Res 2012;106(2-3):71-76.

25 St. George TD, Kirkland PD: Diseases caused by Akabane and related Simbu-group viruses. In: Coetzer J A W, Tustin R C (eds), Infectious Diseases of Livestock, 2nd ed Oxford, UK: Oxford University Press. 2004:1029-36.

26 Radostits OM, Gay CC, Hinchcliff KW, Constable PD: Congenital arthrogryposis and hydranencephaly; Akabane disease; Cache Valley virus disease. Veterinary Medicine: A textbook of the diseases of cattle, horses, sheep, pigs and goats, 10th edition, Saunders Ltd 2007:1207-1209.

27 Zentis HJ, Zentis S, Stram Y, Bernstein M, Rotenberg D, Brenner J: Schmallenberg virus: lessons from related viruses. Vet Rec 2012;171(8):201.

28 Konno S, Moriwaki M, Nakagawa M: Akabane disease in cattle: congenital abnormalities caused by viral infection. Spontaneous disease. Vet Pathol 1982;19(3):246-266.

29 Edwards JF, Livingston CW, Chung SI, Collisson EC: Ovine arthrogryposis and central nervous system malformations associated with in utero Cache Valley virus infection: spontaneous disease. Vet Pathol 1989;26(1):33-39.

30 Uchinuno Y, Noda Y, Ishibashi K, Nagasue S, Shirakawa H, Nagano M, et al: Isolation of Aino virus from an aborted bovine fetus. J Vet Med Sci 1998;60(10):1139-1140.

31 Noda Y, Uchinuno Y, Shirakawa H, Nagasue S, Nagano N, Ohe R, et al: Aino virus antigen in brain lesions of a naturally aborted bovine fetus. Vet Pathol 1998;35(5):409-411.

32 Inaba Y, Kurogi H, Omori T: Letter: Akabane disease: epizootic abortion, premature birth, stillbirth and congenital arthrogryposis-hydranencephaly in cattle, sheep and goats caused by Akabane virus. Aust Vet J 1975;51(12):584-585.

33 Hashiguchi Y, Nanba K, Kumagai T: Congenital abnormalities in newborn lambs following Akabane virus infection in pregnant ewes. Natl Inst of Anim Health Q 1979;19(1-2):1-11.

34 Parsonson IM, Della-Porta AJ, Snowdon WA: Congenital abnormalities in newborn lambs after infection of pregnant sheep with Akabane virus. Infect Immun 1977;15(1):254-262.

35 Parsonson IM, McPhee DA, Della-Porta AJ, McClure S, McCullagh P: Transmission of Akabane virus from the ewe to the early fetus (32 to 53 days). J Comp Pathol 1988;99(2):215-227.

36 Kirkland PD, Barry RD, Harper PA, Zelski RZ: The development of Akabane virus-induced congenital abnormalities in cattle. Vet Rec 1988;122(24):582-586.

37 Kurogi H, Inaba Y, Takahashi E, Sato K, Akashi H, Satoda K, et al: An attenuated strain of Akabane virus: a candidate for live virus vaccine. Natl Inst Anim Health Q 1979;19(1-2):12-22.

38 Kurogi H, Inaba Y, Takahashi E, Sato K, Goto Y: Experimental infection of pregnant goats with Akabane virus. Natl Inst of Anim Health Q 1977;17(1):1-9.

39 Anonymous: SBV cases increase in cattle. Vet Rec 2012;170(11):273.

40 Anonymous: Geographical spread of UK cases of SBV 'within expectations'. Vet Rec 2012;170(9):217.

41 Anonymous: Schmallenberg virus confirmed in south-east England. Vet Rec 2012;170(15):383-386.

42 van den Brom R, Luttikholt SJ, Lievaart-Peterson K, Peperkamp NH, Mars MH, van der Poel WH, et al: Epizootic of ovine congenital malformations associated with Schmallenberg virus infection. Tijdschr Diergeneeskd 2012;137(2):106-111.

43 Anonymous: Findings in cases of SBV infection in sheep in the south-east. Vet Rec 2012;170(20):510-513.

44 Anonymous: Bovine fetal abnormalities continue to be detected in June. Vet Rec 2012;171(10):240-243.

45 Peperkamp K, Dijkman R, van Maanen C, Vos J, Wouda W, Holzhauer M, et al: Polioencephalo- myelitis in a calf due to infection with Schmallenberg virus. Vet Rec 2012;170(22):570.

46 Inaba Y, Matumoto M: Chapter 43: Akabane Virus. In: Dinter S, Morien B (eds), Virus infections of vertebrates, Vol III Virus infections of ruminants Elsevier Science Publishers BV, Amsterdam. 1990, pp 467-80.

47 Loeffen W, Quak S, De Boer-Luijtze E, Hulst M, Van Der Poel W, Bouwstra R, et al: Development of a virus neutralisation test to detect antibodies against Schmallenberg virus and serological results in suspect and infected herds. Acta Vet Scand 2012;54(1):44.

48 Garigliany MM, Bayrou C, Kleijnen D, Cassart D, Desmecht D: Schmallenberg virus in domestic cattle, belgium, 2012. Emerg Infect Dis 2012;18(9):1512-1514.

49 Elbers AR, Loeffen WL, Quak S, de Boer-Luijtze E, van der Spek AN, Bouwstra R, et al: Seroprevalence of Schmallenberg virus antibodies among dairy cattle, the Netherlands, Winter 2011-2012. Emerg Infect Dis 2012;18(7):1065-1071.

50 Ducomble T, Wilking H, Stark K, Takla A, Askar M, Schaade L, et al: Lack of evidence for Schmallenberg virus infection in highly exposed persons, Germany, 2012. Emerg Infect Dis 2012;18(8):1333-1335.

51 ProMED-mail: Schmallenberg virus - Europe (40): no risk to humans, EU, Archive Number: 20120511.1130349. http://www.promedmail.org [posted 11 May 2012].

52 Bilk S, Schulze C, Fischer M, Beer M, Hlinak A, Hoffmann B: Organ distribution of Schmallenberg virus RNA in malformed newborns. Vet Microbiol 2012;159(1-2):236-238.

53 Anonymous: Serological ELISA for SBV currently being evaluated. Vet Rec 2012;170(18):453.

54 Kurogi H, Inaba Y, Takahashi E, Sato K, Omori T, Miura Y, et al: Epizootic congenital arthrogryposis-hydranencephaly syndrome in cattle: isolation of Akabane virus from affected fetuses. Arch Virol 1976;51(1-2):67-74.

55 De Regge N, Van Loo H, Bertels G, Saulmont M, Vandenbussche F, van den Berg T, et al: Comparative analysis to determine the most appropriate tissue for Schmallenberg virus detection by RT-PCR in aborted lambs and calves. Proceedings of the EPIZONE 6th Annual Meeting Brighton, June, 2012. 2012:202.

56 OIE: OIE technical factsheet on Schmallenberg virus. Available from: http://www.oie.int/en/our-scientific-expertise/specific-information-and-recommendations/schmallenberg-virus/2012.

57 van Maanen C, van der Heijden H, Wellenberg GJ, Witteveen G, Luttikholt S, Bouwstra R, et al: Schmallenberg virus antibodies in bovine and ovine fetuses. Vet Rec 2012;171(12):299.

58 Poel WH: Diagnostics for Schmallenberg virus. Vet Rec 2012;171(12):294-295.

59 Anonymous: Fetal deformities detected during surveillance for Schmallenberg virus. Vet Rec 2012;171(5):117-120.

60 Parsonson IM, Della-Porta AJ, O'Halloran ML, Snowdon WA, Fahey KJ, Standfast HA: Akabane virus infection in the pregnant ewe: 1. Growth of virus in the foetus and the development of the foetal immun response. Vet Microbiol 1981;6:197-207.

61 Charles JA: Akabane virus. Vet Clin North Am Food Anim Pract 1994;10(3):525-546.

62 Kurogi H, Inaba Y, Goto Y, Miura Y, Takahashi H. Serologic evidence for etiologic role of Akabane virus in epizootic abortion-arthrogryposis-hydranencephaly in cattle in Japan, 1972-1974. Arch Virol 1975;47(1):71-83.

63 Höper D, Wernike K, Eschbaumer M, Conraths F, Hoffmann B, Schirrmeier H, et al: Schmallenberg-Virus, ein neues Virus in Europa. Dtsch Tierärtebl. 2012;04/2012:500-5.

Martin Beer, Institute of Diagnostic Virology, Friedrich-Loeffler-Institut, Suedufer 10, 17493 Greifswald–Insel Riems, Germany

Tel: +49 38351 71200 , fax: +49 38351 71275

Email: martin.beer@fli.bund.de

Roth JA, Richt JA, Morozov IA (eds): Vaccines and Diagnostics for Transboundary Animal Diseases. Dev Biol (Basel). Basel, Karger, 2013, vol 135, pp 183-189.

Q Fever Diagnosis and Control in Domestic Ruminants

H.I.J. Roest[1], *A. Bossers*[2], *J.M.J. Rebel*[2]

[1] Department of Bacteriology and TSEs, Central Veterinary Institute, part of Wageningen UR, Lelystad, The Netherlands

[2] Department of Infection Biology, Central Veterinary Institute, part of Wageningen UR, Lelystad, the Netherlands

Key words: Q fever, *Coxiella burnetii*, diagnosis, control, vaccination, vaccine, domestic ruminants

Abstract: Q fever is a zoonosis caused by the bacterium *Coxiella burnetii*, a highly infectious agent that can survive in the environment. Therefore, Q fever has a major public health impact when outbreaks occur. Small ruminants are identified as the source in the majority of outbreaks in humans. Accurate diagnosis and effective control strategies are necessary to limit the zoonotic and veterinary impact of Q fever. For this, knowledge of the pathogenesis of Q fever and excretion routes of *C. burnetii* from infected animals is crucial. Abortions as well as normal parturitions in infected small ruminants are the most important excretion routes of *C. burnetii*. Excretion of *C. burnetii* via faeces and vaginal mucus has also been suggested. However, contamination of these samples by bacteria present in the environment may influence the results. This hampers the accurate identification of infected animals by these samples; however, the detection of *C. burnetii* in milk samples seems not to be influenced by environmental contamination. Q fever in animals can be detected by direct (immunohistochemistry and PCR) and indirect (complement fixation test (CFT), enzyme-linked immunosorbent assay (ELISA) and indirect immunofluorescence assay (IFA) methods. A combination of both direct and indirect methods is recommended in current protocols to detect Q fever on herd level. For the control of Q fever in domestic animals, vaccination with a phase 1 *C. burnetii* whole cell inactivated vaccine is reported to be effective in preventing abortion and reducing bacterial shedding, especially after several years of administration. Vaccination might not be effective in already infected animals nor in pregnant animals. Furthermore, the complicated vaccine production process, requiring biosafety level 3 facilities, could hamper vaccine availability. Future challenges include the development of improved, easier to produce Q fever vaccines.

INTRODUCTION

Q fever, a zoonosis caused by the intracellular bacterium *Coxiella burnetii*, has a worldwide distribution, except for New Zealand. Due to its low minimal infectious dose and high transmissibility, *C. burnetii* is considered a biothreat agent [1]. Unique properties of the bacterium are its developmental cycle and lipopolysaccharide (LPS) phase variation. In the developmental cycle of *C. burnetii*, two distinct variants have been identified: a large cell variant (LCV) and a small cell variant (SCV). The LCV is the metabolically active and replicative entity of the bacterium. The SCV, which is the spore-like form of *C. burnetii*, is highly resistant to environmental stress, such as high temperatures, UV radiation and osmotic pressure. The SCV allows the bacterium to survive in the environment while keeping its infectivity [2]. Two phases of the *Coxiella*'s LPS are recognised: the phase 1 phenotype expresses full length LPS on its surface and is highly virulent, while in phase 2 the LPS is truncated, lacking the O-antigenic region. This phase variation appears to have implications for vaccines as a phase 1 *C. burnetii* vaccine is reported to be effective in goats, but a phase 2 vaccine is not effective [3].

C. burnetii can infect a wide range of hosts including humans, arthropods, birds, terrestrial and marine mammals. Humans usually acquire Q fever after inhalation of environmentally present *C. burnetii*; however, almost 60% of human infections are asymptomatic. Among the 40% symptomatic acute Q fever patients, the majority will present a non-specific, self-limiting illness. More severe clinical symptoms include fever, headache, atypical pneumonia and hepatitis, with one to five per cent of patients developing chronic Q fever, which may result in life-threatening endocarditis. Domestic ruminants are the main reservoir for human Q fever; and the most important clinical presentations of Q fever in these animals are abortion and stillbirth. Furthermore, high numbers of bacteria are excreted during abortion or parturition and spread into the environment. In cattle, *C. burnetii* infection has a less dramatic clinical appearance and is mainly associated with metritis as well as with reproduction problems.

Clearly, Q fever has a major public health impact when outbreaks occur; and outbreaks in humans are reported frequently and worldwide, involving up to 415 laboratory confirmed human cases per outbreak; however, outbreaks are normally localised and restricted to one episode. Nevertheless, the Dutch Q fever outbreak occurred for four successive years (2007-2010) [4]. Sheep have been identified as the source in the majority of outbreaks, with goats as 'second best'; and these two species have been held responsible for the biggest outbreaks reported. Remarkably, human outbreaks due to infected sheep have also been reported after normal lambing [5, 6], but only a few outbreaks in humans have been related to cattle and cats [7].

To limit the zoonotic and veterinary impact of Q fever, accurate diagnosis and effective control strategies are necessary. The present review provides an update on diagnosis and control on Q fever in domestic ruminants, including recent outcomes on the excretion of *C. burnetii*.

EXCRETION OF C. *BURNETII*

Knowledge of the pathogenesis and excretion of C. *burnetii* from infected animals is crucial for the accurate diagnosis and understanding of both transmission routes and the risks of human infection. Abortions in C. *burnetii* infected domestic ruminants are accompanied by massive excretion of bacteria which spread into the environment. This seems to be the most important excretion route of C. *burnetii*, as up to 10^9 organisms per gram placenta tissue are excreted [8]. It has been suggested that, with the birth of healthy calves, kids or lambs from infected animals, fewer bacteria are excreted compared to abortions [9]. However, this may not hold true. In infected sheep, normal parturitions have been the most probable source of several Q fever outbreaks [5, 6]. In a recent experimental study, we showed that infected goats gave birth to both live and dead kids. High numbers of C. *burnetii* were excreted not only during abortion, but also during parturition of live born kids [10]. These results imply that normal parturitions, like abortions, in infected small ruminants should be considered as important excretion routes of C. *burnetii.*

C. *burnetii* has also been detected in faeces, vaginal mucus and milk of infected domestic ruminants [8, 11-13]. For example, in goat herds, in both aborting and non-aborting goats, C. *burnetii* DNA has been detected in faeces, vaginal mucus and milk [13]. Also in cattle, variable excretion via these three routes has been reported, sometimes independent of an abortion history. Sixty-five per cent of cows seem to shed C. *burnetii* evenly distributed over one of the three routes; however, cows that excrete C. *burnetii* by all three routes are scarce [12]. Results of recent experimental infections in goats could give an explanation for the variable excretion of the agent in bodily fluids [10]. No active replication of C. *burnetii* in the gut, liver or female genital tract (besides the trophoblasts) of goats could be detected before or after parturition [10, 14]. This could indicate that active excretion of the Q fever bacterium does not occur. However, C. *burnetii* DNA has been detected in faeces and vaginal mucus, but only after the first kidding in the group. This excretion continued after the disappearance of C. *burnetii* DNA from the goat's body [10]. As suggested earlier [15], we assume that detection of C. *burnetii* DNA in faeces and vaginal mucus before and after parturition is due to contamination with C. *burnetii* present in the environment. Thus the presence of positive faecal and vaginal samples reflects more the level of C. *burnetii* contamination in the sampling environment than the actual infection status of the animal. Therefore, faecal and vaginal mucus samples should be regarded as inadequate to determine the infection status of an individual animal.

In contrast to faecal and vaginal samples, the presence of C. *burnetii* in milk samples seems not to be influenced by environmental contamination. The excretion of C. *burnetii* stopped with the disappearance of C. *burnetii* DNA from the mammary gland [10]. Therefore, aseptically taken milk samples can be considered the best routinely collectible samples reflecting the individual infection status.

Diagnosis of Q Fever in Domestic Ruminants

The occurrence of abortion is not confined to Q fever. Therefore, evidence of a C. *burnetii* infection requires laboratory confirmation. C. *burnetii* can be detected in infected placenta material via immunohistochemistry (IHC) using C. *burnetii* specific antibodies. When C. *burnetii* is the cause of abortion, high numbers of bacteria are usually observed in the trophoblasts of the allantochorion [10, 14]. The allantochorion and the base of the cotyledon appear to be the best locations to detect Q fever bacterium in an infected placenta [10].

C. *burnetii* can also be detected with specific polymerase chain reaction (PCR) tests. Specific and sensitive detection of C. *burnetii* can be achieved with primers and probes targeting the multi-copy IS1111 genetic element, because this C. *burnetii*-specific multiple copy element allows greater sensitivity detection compared to primers and probes that target single copy genes [16]. Furthermore, PCR can be used to detect C. *burnetii* DNA in a wide range of samples, including placenta tissues, faeces, vaginal mucus and milk [17].

Apart from histopathology and PCR, several indirect diagnostic methods of Q fever detection are available. C. *burnetii* specific antibodies in serum can be measured using the complement fixation test (CFT), the enzyme-linked immunosorbent assay (ELISA) or the indirect immunofluorescence assay (IFA) test. The CFT is most widely used in veterinary laboratories, although commercially available ELISAs to detect C. burnetii antibodies in ruminants are increasingly used. The IFA is less widely used as this test is not commercially available for veterinary use. Compared to ELISA, the CFT lacks sensitivity to detect C. *burnetii* specific antibodies in ruminant serum [7, 18, 19], although contrary findings have also been reported [20]. Several reports indicate that C. *burnetii* can be shed by cattle, sheep and goats without a detectable serological response. Although this could be due to the delay in the serological response after a primary infection, it has been suggested that serology is not a useful tool to diagnose abortions at the individual level [11, 13, 21]. However, this suggestion is based on the detection of C. *burnetii* in faeces, vaginal mucus or milk in individuals without a serological response. As discussed above, this may not be fully justified, since misclassification of infected animals can easily occur when using faecal and vaginal samples, due to environmental contamination [10].

Current protocols to detect Q fever at herd level recommend a combination of both direct and indirect techniques to diagnose Q fever in cattle and small ruminant herds. It has been recommended that samples from abortion/parturition material and/or from vaginal mucus of two to six animals of the herd are investigated by PCR within eight days after abortion or parturition. In addition, serum samples from at least six to ten animals should be investigated, especially when PCR results are negative. These animals should preferably have delivered at least 15 days prior to sampling [22]. Aseptically taken milk samples have an additional value in the detection of C. *burnetii* infection; and to detect a Coxiella infection at herd level bulk tank milk samples could have additional value [23].

CONTROL OF Q FEVER IN DOMESTIC RUMINANTS

To control Q fever in domestic ruminants several methods are available. Stamping out of pregnant goats on Q fever infected farms was used in the Dutch Q fever outbreak [24]. Two less drastic methods of control are antibiotic treatment and/or vaccination. In humans, acute and chronic Q fever is treated with antibiotics like doxycycline and hydrochloroquine. Although thorough evaluations of these therapies are lacking, limited evidence suggests these therapies are effective. The effect of antibiotics to treat Q fever in animals has been poorly studied. Although a few studies suggest a reduction in the abortion rate and in the number of bacteria excreted at parturition [25, 26], this effect could not be observed in small ruminants in the Netherlands treated with oxytetracycline [27]. Nevertheless, in dairy cattle a beneficial effect of antibiotic treatment complementary to vaccination with a phase 1 C. *burnetii* inactivated vaccine was observed [28]. Because the effect of antibiotic treatment is not clearly evidence-based, infected animals are difficult to detect, and prudent use of antibiotics is needed to avoid microbial resistance. Therefore, on balance, Q fever in animals should not be treated with antibiotics.

Vaccination of animals with a phase 1 C. *burnetii* whole cell inactivated vaccine has been reported to be effective in preventing abortion, as well as in reducing bacterial shedding in goats and cattle [3, 29, 30]. Vaccination in goats and cattle is most effective in primiparous animals and reduces both the prevalence of shedders, as well as the bacterial load [29-31]. However, vaccination might not be effective in goats or cattle that have been infected before vaccination nor in pregnant animals [29, 31]. Probably, several successive years of vaccination are needed to control Q fever at herd level [32]. These results indicate that Q fever in domestic ruminants can best be prevented by vaccination. In the Netherlands, Q fever vaccination of dairy goats and dairy sheep with a phase 1 C. *burnetii* vaccine has been used as one of the measures to reduce the bacterial shedding in small ruminants to control Dutch Q fever outbreaks [24].

Efficacy of Q fever vaccines seems to be related to the LPS phase of the *Coxiella* incorporated in the vaccine. The current LPS phase 2 C. burnetii vaccine has not been effective in contrast to vaccination with a LPS phase 1 C. *burnetii* vaccine [33], which has been effective. Therefore, it has been suggested that the phase 1 LPS plays a key role in the vaccine induced protection [34]. Cross immunity among strains of C. *burnetii* have shown to be complete in experimentally infected animals [35, 36]. Furthermore, field observations confirmed this, as the Nine Mile strain derived phase 1 C. *burnetii* vaccine was effective in the Dutch Q fever outbreak. A genotype distinct from the Nine Mile strain also played a key role in this outbreak [30, 37]. The main drawbacks of the commercially available phase 1 C. *burnetii* whole cell inactivated vaccine are the complicated production process and the need for dedicated biosafety level 3 facilities. This hampers vaccine availability when demand increases. Easier to produce vaccines (e.g. recombinants vaccines) could well overcome these drawbacks, but are not yet available.

Remaining Questions

Remaining questions in the diagnosis and control of Q fever in domestic ruminants include the detection of infected pregnant animals as they do not shed the bacterium, as well as the role of non-pregnant animals in the epidemiology of Q fever. With respect to the control of Q fever, major challenges are to differentiate between naturally infected and vaccinated animals, as well as to increase vaccination efficacy in pregnant or already infected animals by improved easier to produce Q fever vaccines.

References

1 Oyston PC, Davies C: Q fever: the neglected biothreat agent. J Med Microbiol 2011;60(Pt 1):9-21.

2 McCaul TF, Williams JC: Developmental cycle of *Coxiella burnetii*: Structure and morphogenesis of vegetative and sporogenic differentiations. J Bacteriol 1981;147(3):1063-1076.

3 Arricau-Bouvery N, Souriau A, Bodier C, Dufour P, Rousset E, Rodolakis A: Effect of vaccination with phase I and phase II *Coxiella burnetii* vaccines in pregnant goats. Vaccine 2005;23(35):4392-4402.

4 Dijkstra F, van der Hoek W, Wijers N, Schimmer B, Rietveld A, Wijkmans CJ, et al: The 2007-2010 Q fever epidemic in The Netherlands: Characteristics of notified acute Q fever patients and the association with dairy goat farming. FEMS Immunol Med Microbiol 2012;64(1):3-12.

5 Koene RP, Schimmer B, Rensen H, Biesheuvel M, A DEB, Lohuis A, et al: A Q fever outbreak in a psychiatric care institution in The Netherlands. Epidemiol Infect 2010;9:1-6.

6 Porten K, Rissland J, Tigges A, Broll S, Hopp W, Lunemann M, et al: A super-spreading ewe infects hundreds with Q fever at a farmers' market in Germany. BMC Infect Dis 2006;6:147.

7 EFSA: Scientific Opinion on Q fever. EFSA Journal [serial on the Internet] 2010;8(5): Available from: http://www.efsa.europa.eu/en/scdocs/doc/1595.pdf.

8 Arricau Bouvery N, Souriau A, Lechopier P, Rodolakis A: Experimental *Coxiella burnetii* infection in pregnant goats: Excretion routes. Vet Res 2003; 34(4):423-433.

9 Rodolakis A: Q Fever in dairy animals. Ann NY Acad Sci 2009;1166:90-93.

10 Roest HJ, van Gelderen B, Dinkla A, Frangoulidis D, van Zijderveld FG, Rebel J, et al: Q fever in pregnant goats: pathogenesis and excretion of *Coxiella burnetii*. PLoS One 2012;7(11):14.

11 Berri M, Souriau A, Crosby M, Crochet D, Lechopier P, Rodolakis A: Relationships between the shedding of *Coxiella burnetii*, clinical signs and serological responses of 34 sheep. Vet Rec 2001;148(16):502-505.

12 Guatteo R, Beaudeau F, Berri M, Rodolakis A, Joly A, Seegers H: Shedding routes of *Coxiella burnetii* in dairy cows: implications for detection and control. Vet Res 2006;37(6):827-833.

13 Rousset E, Berri M, Durand B, Dufour P, Prigent M, Delcroix T, et al: *Coxiella burnetii* shedding routes and antibody response after outbreaks of Q fever-induced abortion in dairy goat herds. Appl Environ Microbiol 2009 ;75(2):428-433.

14 Sanchez J, Souriau A, Buendia AJ, Arricau-Bouvery N, Martinez CM, Salinas J, et al: Experimental *Coxiella burnetii* infection in pregnant goats: A histopathological and immunohistochemical study. J Comp Pathol 2006;135(2-3):108-115.

15 Welsh HH, Lennette EH, Abinanti FR, Winn JF: Air-borne transmission of Q fever: the role of parturition in the generation of infective aerosols. Ann NY Acad Sci 1958;70(3):528-540.

16 Klee SR, Tyczka J, Ellerbrok H, Franz T, Linke S, Baljer G, et al: Highly sensitive real-time PCR for specific detection and quantification of *Coxiella burnetii*. BMC Microbiol 2006;6:2.

17 OIE Manual of Diagnostic Tests and Vaccines for Terrestrial Animals 2012. http://www.oie.int/fileadmin/Home/eng/Health_standards/tahm/2.01.12_Q-FEVER.pdf, accessed 30 July 2012

18 Horigan MW, Bell MM, Pollard TR, Sayers AR, Pritchard GC: Q fever diagnosis in domestic ruminants: comparison between complement fixation and commercial enzyme-linked immunosorbent assays. J Vet Diagn Invest 2011;23(5):924-931.

19 Rousset E, Durand B, Berri M, Dufour P, Prigent M, Russo P, et al: Comparative diagnostic potential of three serological tests for abortive Q fever in goat herds. Vet Microbiol 2007;124(3-4):286-297.

20 Emery MP, Ostlund EN, Schmitt BJ: Comparison of Q fever serology methods in cattle, goats, and sheep. J Vet Diagn Invest 2012;24(2):379-382.

21 Guatteo R, Joly A, Beaudeau F: Shedding and serological patterns of dairy cows following abortions associated with *Coxiella burnetii* DNA detection. Vet Microbiol 2012;155(2-4):430-433.

22 Rousset E, Durand B, Champion JL, Prigent M, Dufour P, Forfait C, et al: Efficiency of a phase 1 vaccine for the reduction of vaginal Coxiella burnetii shedding in a clinically affected goat herd. Clin Microbiol Infect 2009;15(s2):188-189.

23 van den Brom R, van Engelen E, Luttikholt S, Moll L, van Maanen K, Vellema P: *Coxiella burnetii* in bulk tank milk samples from dairy goat and dairy sheep farms in The Netherlands in 2008. Vet Rec 2012;170(12):310.

24 Roest HI, Tilburg JJ, van der Hoek W, Vellema P, van Zijderveld FG, Klaassen CH, et al: The Q fever epidemic in The Netherlands: History, onset, response and reflection. Epidemiol Infect 2011;139(1):1-12.

25 Berri M, Crochet D, Santiago S, Rodolakis A: Spread of *Coxiella burnetii* infection in a flock of sheep after an episode of Q fever. Vet Rec 2005;157(23):737-740.

26 Muskens J, Mars MH, Franken P: [Q fever: an overview]. Tijdschr Diergeneeskd. 2007;132(23):912-917.

27 Wouda W, Dercksen DP:[Abortion and stillbirth among dairy goats as a consequence of *Coxiella burnetii*]. Tijdschr Diergeneeskd. 2007;132(23):908-911.

28 Taurel AF, Guatteo R, Joly A, Beaudeau F: Effectiveness of vaccination and antibiotics to control *Coxiella burnetii* shedding around calving in dairy cows. Vet Microbiol 2012;159(3-4):432-437.

29 Guatteo R, Seegers H, Joly A, Beaudeau F: Prevention of *Coxiella burnetii* shedding in infected dairy herds using a phase I *C. burnetii* inactivated vaccine. Vaccine 2008;26(34):4320-4328.

30 Hogerwerf L, van den Brom R, Roest HI, Bouma A, Vellema P, Pieterse M, et al: Reduction of *Coxiella burnetii* prevalence by vaccination of goats and sheep, The Netherlands. Emerg Infect Dis 2011;17(3):379-386.

31 Rousset E, Durand B, Champion JL, Prigent M, Dufour P, Forfait C, et al: Efficiency of a phase 1 vaccine for the reduction of vaginal *Coxiella burnetii* shedding in a clinically affected goat herd. Clin Microbiol Infect 2009;15 Suppl 2:188-189.

32 Astobiza I, Barandika JF, Ruiz-Fons F, Hurtado A, Povedano I, Juste RA, et al: Four-year evaluation of the effect of vaccination against *Coxiella burnetii* on reduction of animal infection and environmental contamination in a naturally infected dairy sheep flock. Appl Environ Microbiol 2011;77(20):7405-7407.

33 Arricau-Bouvery N, Rodolakis A: Is Q fever an emerging or re-emerging zoonosis? Vet Res 2005;36(3):327-349.

34 Zhang G, Russell-Lodrigue KE, Andoh M, Zhang Y, Hendrix LR, Samuel JE: Mechanisms of vaccine-induced protective immunity against *Coxiella burnetii* infection in BALB/c mice. J Immunol 2007;179(12):8372-8380.

35 Arricau-Bouvery N, Hauck Y, Bejaoui A, Frangoulidis D, Bodier CC, Souriau A, et al: Molecular characterization of *Coxiella burnetii* isolates by infrequent restriction site-PCR and MLVA typing. BMC Microbiol. 2006;6:38.

36 Ormsbee RA, Bell BP, Lackman DB, Tallent GL: The influence of phase on the protective potency of Q fever vaccine. J Immunol 1964(92):404-412.

37 Roest HI, Ruuls RC, Tilburg JJ, Nabuurs-Franssen MH, Klaassen CH, Vellema P, et al: Molecular Epidemiology of *Coxiella burnetii* from Ruminants in Q Fever Outbreak, the Netherlands. Emerg Infect Dis 2011 ;17(4):668-675.

Hendrik-Jan Roest, Edelhertweg 15, 8219 PH Lelystad, The Netherlands
Tel: +31 320 238026, fax: +31 320 238153
Email: hendrikjan.roest@wur.nl

Roth JA, Richt JA, Morozov IA (eds): Vaccines and Diagnostics for Transboundary Animal Diseases. Dev Biol (Basel). Basel, Karger, 2013, vol 135, pp 191-200.

Opportunities in Diagnostic and Vaccine Approaches to Mitigate Potential Heartwater Spreading and Impact on the American Mainland

N. Vachiéry[1*], *I. Marcelino*[2,3], *D. Martinez*[1], *T. Lefrançois*[4]

[1] CIRAD, UMR CMAEE, Petit-Bourg, Guadeloupe, France
[2] IBET, Oeiras, Portugal
[3] ITQB-UNL, Oeiras, Portugal
[4] CIRAD, UMR CMAEE, Montpellier, France

Key words: diagnostic, vaccine, heartwater control

Abstract: Heartwater, caused by the *Rickettsiales Ehrlichia ruminantium* (ER), is a tropical tick-borne disease of wild and domestic ruminants, transmitted by *Amblyomma* ticks. It causes significant economic losses due to high mortality and the high cost of antibiotic treatment of affected animals, limiting herd productivity. It is present in sub-Saharan Africa, islands in the Indian Ocean and two Caribbean islands (Guadeloupe and Antigua) from where it threatens the American mainland due to risk of the spread of infected *A. variegatum* by migratory birds or by uncontrolled movement of animals. If an accidental introduction of a tick-free ER carrier animal occurs, autochthonous *A. maculatum* has proven to be a good experimental vector for heartwater. Modeling *A. variegatum* population dynamics has been developed, but further work is needed to predict favourable habitats and allow targeted surveillance. We overview here the advances in diagnostics, vaccines and epidemiology of heartwater and analyze the research gaps and needs to mitigate potential ER introduction and spread on the American mainland. Effective serologic ELISA tests allow prevalence studies, and several PCR-based diagnostic tests are currently available to detect ER in sick animals. However, the development of rapid assays, including multi-pathogen tests, would enhance the efficacy and cost-effectiveness of heartwater diagnosis. Several experimental vaccines (inactivated, attenuated and recombinant) are under development. Attenuated and inactivated vaccines are effective against homologous strains but their efficacy in the field is decreased due to broad antigenic diversity of ER. New molecular typing assays are now being used to study the genetic structure of ER populations worldwide, but the linking of genotyping to cross-protection is still not straightforward. Currently an inactivated vaccine would be the most appropriate vaccine for the American mainland due to its safety, the availability of a fully controlled bioprocess allowing ER mass production and the possibility to design "regional cocktail vaccines". This would require the selection and isolation of Caribbean ER strains supported by data of molecular epidemiology studies in this region. Development of an universal recombinant vaccine requires increased knowledge of ER biology, including virulence mechanisms. Comparison of virulent and attenuated strains using "omic approaches" is on-going and will be crucial to understand these mechanisms and to develop improved vaccines.

INTRODUCTION

Disease and causal agent

Heartwater is a fatal tropical tick-borne disease of wild and domestic ruminants. The causal agent is *Ehrlichia ruminantium* (ER), an obligate intracellular α-proteobacterium belonging to the order *Rickettsiales* and the *Anaplasmataceae* family. The *Anaplasmataceae* family contains 4 genera: *Anaplasma*, *Wolbachia*, *Neorickettsia* and *Ehrlichia*. Based on 16S sequence, *Ehrlichia ruminantium* is closely related to *E. chaffeensis, ewingii, muris, ovis* and *canis* [1]. Heartwater is transmitted by *Amblyomma* ticks, mainly *A. variegatum* (present in Africa and in the Caribbean islands) and *A. hebraeum* (present in Southern Africa).

Heartwater mortality rates depend on the virulence of ER strains and host species. Small ruminants (caprine and ovine) are more susceptible than cattle and there is a high susceptibility of imported high production breeds, avoiding any improvement of productivity in the herds in Africa and in Caribbean islands. The economic impact due to mortality and cost of antibiotic and acaricide treatments was estimated as $47.6 million per year for the Southern Africa community (Angola, Botswana, Malawi, Mozambique, South Africa, Swaziland, Tanzania and Zimbabwe).

Heartwater epidemiology

Heartwater is present in sub-Saharan Africa, the Indian Ocean (Madagascar, Comoros islands, Mayotte, Reunion Island, Mauritius Island) and two Caribbean islands (Guadeloupe and Antigua). *A. variegatum* was introduced from West Africa into Guadeloupe and Martinique at the end of the 18th century and spread to Antigua in the 19th century. Further expansion to other Lesser Antilles islands occurred at mid-20th century and coincided with the spread of the cattle egret, a cosmopolitan species of heron.

To control tick spreading in the Caribbean, several eradication programmes have been created. From 1994 to 2008, the Caribbean *Amblyomma* program (CAP) was implemented in the English Lesser Antilles (Anguilla, Antigua, Barbuda, Barbados, Dominica, Montserrat, Nevis, St Kitts, St Lucia and Saint Maarten) to eradicate *Amblyomma* ticks. By the end of the project, a decrease in tick infestation was observed on six islands: Saint Lucia, Saint Kitts, Montserrat, Anguilla, Dominica and Barbados. Nowadays, two islands have rare *Amblyomma* ticks (Saint Vincent and Saint Croix) and five islands are *Amblyomma* free (Anguilla, Barbuda, Montserrat, Grenada and Barbados), as shown in Figure 1. In Dominica and Saint Lucia hot spots are currently observed, whereas Martinique, Antigua, Saint Marteen, Saint Kitts and Nevis have low to moderate levels of tick infestation. Another control program targeting French islands against *A. variegatum* tick was also put into practice, but the level of infestation remained high in Guadeloupe and Marie Galante. Recent studies completed on these two islands indicate that the prevalence of tick infested herds is 35.6% and 73.8% in Guadeloupe and Marie Galante, respectively, with 36.7% and 19.1% of ER infection rates [2, 3]. Therefore, these two islands constitute a reservoir for ticks and ER in the Caribbean which poses a continuing threat to the American mainland due to potential introduction of infected *A. variegatum* through migratory birds or uncontrolled movement of animals [4]. If an accidental introduction of a tick-free ER carrier animal does occur, autochthonous *A. maculatum* and *A. cajenense* (which have proven to be experimental vectors for heartwater) can promote ER spreading from the North of Mexico down to the South of Brazil (except for the Andean region).

Although eradication programmes are now on standby in the Caribbean, heartwater control and early diagnostics are a major concern in Guadeloupe. A surveillance network (RESPANG) was recently created to detect sick animals with heartwater and, in case of positive detection, to perform a sensitization campaign.

Modeling *A. variegatum* population dynamics

In order to improve tick control, a model to study *A. variegatum* population dynamics and further predict potential ecological niches was developed using biological data from field and experimental conditions, including climatic parameters (temperature, humidity, photo period) [5]. The areas of putative establishment of *A. variegatum* were defined, including the Gulf of Mexico and part of South America; and the potential for the introduction and establishment of *A. variegatum* and ER was confirmed. Applications of this model are not restricted to habitat suitability; they also include evaluation of control and surveillance strategies. This model will be useful in case of disease introduction to the American mainland.

To control the threat of heartwater outbreaks, early detection and prevention of the disease are required. Below, we will overview the current advances on diagnostics, vaccines and epidemiology of heartwater and analyze the research gaps and the essential need to control ER introduction and spread on the American mainland.

Serological Diagnostics

Two serological diagnostics based on the detection of antibodies against a major antigenic ER protein of MAP-1 are available: a competitive ELISA MAP1 and an indirect ELISA using a fraction of MAP1 protein, MAP1-B [6, 7]. These assays display cross-reaction with *E. chaffeensis* and *E. muris*. The indirect MAP1-B ELISA is used routinely at the regional OIE reference laboratory. ELISAs against heartwater are suitable for prevalence studies but should not be used neither for diagnostic of clinical cases nor to evaluate the epidemiological status of imported animals. Indeed, there is a 15 days delay in seroconversion after animal infection, and the seropositivity period lasts several weeks for bovines and less than 6 months for small ruminants.

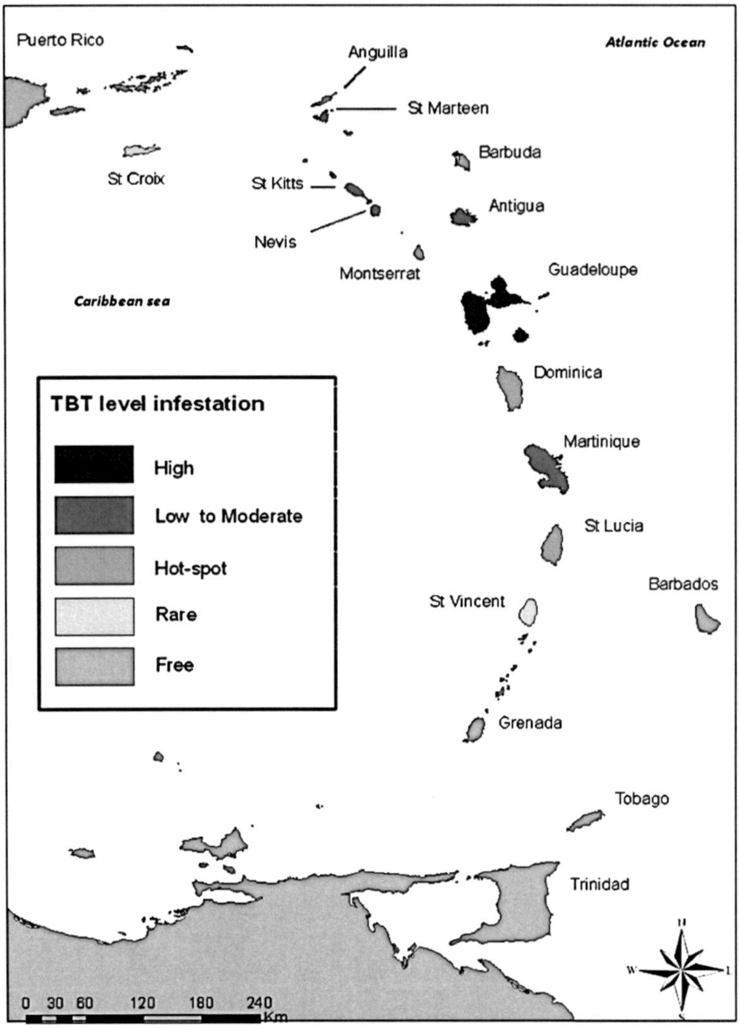

Fig.1. *Amblyomma variegatum* ticks (TBT=tropical bont ticks) infestation in Lesser Antilles, data from CaribVET network (2011).

Molecular Diagnostics

Detection of ER by nested PCR

Nested PCR targeting of a specific fragment of ER gene *pCS20*, is a highly specific and sensitive method with a detection level down to 6 copies of ER per sample, which has been validated for over 100 different ER strains [2, 8]. This method can be used on the DNA matrix from ticks, organs (brain, lung, spleen) from dead animals and blood from sick animals during the hyperthermia onset, using either fresh or preserved samples in ethanol. It is also used in molecular epidemiological studies [2, 9]. The disadvantages of this method include potential cross-contamination and the associated constraint of specific handling and careful organization of the diagnostic laboratory. It is also a time-consuming assay which requires two days before results are obtained.

Detection of ER by real-time PCR

Several real-time PCR assays targeting ER genes are available [10-12]. Real-time PCR targeting *map1* and *map1-1* polymorphic genes were tested on a limited number of strains (up to six), and in these conditions cannot yet be used for diagnostic. Another real-time PCR assay targeting the *pCS20* region has a sensitivity level similar to the nested PCR, but it was tested so far only on 15 different strains, and screening of additional strains should be performed. Furthermore, using real-time PCR would reduce contamination problems and is quicker than nested PCR.

Identification of suitable diagnostic for the American mainland and further needs

A comparison of different molecular diagnostic methods for heartwater and their suitability for use in case of ER spreading to the American mainland is presented in Table 1. Nested PCR targeting the *pCS20* region (PCR *pCS20*) is the gold standard assay for the diagnostic of heartwater in the reference laboratory and is suitable for ER detection in the American mainland in suspicious cases. Compared to nested PCR, real-time PCR *pCS20* is faster and less prone to cross-contamination. It is thus essential to do further validation and to compare the detection efficiency between real-time and nested PCR *pCS20* in order to validate the real-time PCR assay for heartwater diagnosis. A molecular multi-pathogen method, the reverse line blot, is also available, targeting *Anaplasma* and *Ehrlichia* sp. including ER, *Ehrlichia ovina, Anaplasma marginale, A. centrale, A. ovis* and *A. phagocytophilum* [13]. However, this method is less sensitive than nested PCR and has not been validated in large scale epidemiological studies. There is therefore a strong interest in developing improved rapid multi-pathogen detection assays, i.e. using microarray technology or multiplex real-time PCR that could also include other tick-borne pathogens such as *Babesia* and *Theileria* sp.

Table 1: Comparative analysis of molecular diagnostic tests for ER

Critical Criteria	Nested PCR *pCS20*	qPCR *map1*	qPCR *map1-1*	qPCR *pCS20*	Research needs
Speed	5	8	8	8	Development of quick tests: Multi-pathogen detection by microarrays or multiplex qPCR
Sensitivity	8	6	5	8	
Easiness of handling & sensitivity to cross contamination	5	8	8	8	
Detection of multiple ER strains	10	**2**	**2**	**3**	Comparison of detection capability between *pCS20* nested PCR and qPCR
Adaptation to field samples	10	NT	5	5	
Total Score	**38**	24	29	32	

NT = not tested
Scoring from 0 to 10, where 0 = the worst and 10 = the best

Vaccination Approaches

"Infection and treatment" method

The "infection and treatment" method is the only vaccine commercially available, and that only in South Africa. It uses the injection of frozen blood from infected sheep with live virulent ER organisms (Ball3 strain) inoculated intravenously, followed by a treatment with tetracycline during the early hyperthermia period [14]. This method requires a daily monitoring of vaccinated animals and has a high risk of mortality. As it is a live vaccine, there are major practical and safety concerns, such as cold chain requirements, risk of infection by other pathogens and spreading of the live vaccine strain. Therefore, this vaccine is not suitable for America where heartwater is not widespread.

Recombinant vaccine

Several recombinant vaccines based on DNA and the recombinant protein prime/boost approach have been evaluated [14]. The use of 4 ORFs DNA/recombinant proteins prime/boost resulted in a 100% survival rate after homologous needle challenge, but lower survival rates (20% of protection) after heterologous field challenge in South Africa. Recently, a prime/boost vaccination trial using the polymorphic *cpg1* gene and the recombinant protein also resulted in complete protection of vaccinated animals after homologous challenge; no trials with heterologous strains have yet been performed. Due to the polymorphic property of *cpg1*, a cocktail of representative CpG1 from different strains should be included in the vaccine before any field trial. Although recombinant vaccines look promising under experimental conditions, results during field trials have been less successful. Moreover, the use of a gene gun is necessary for efficient DNA vaccination, which is not suitable for large field vaccination and for use on the American mainland.

Attenuated vaccine

Currently, three ER strains (Senegal from West Africa, Welgevonden from South Africa and Gardel from Guadeloupe, French West Indies) have been attenuated by successive in vitro passages; and their efficacy as vaccine candidates has been proven under experimental conditions [14]. They can confer good protection against homologous strains with low morbidity, but only variable protection has been observed against heterologous challenges. So far, the Welgevonden attenuated vaccine strain demonstrated good protection in controlled trials against homologous and four heterologous strains, but has not yet been tested in field conditions. Since they are live vaccines, attenuated vaccine must be stored in liquid nitrogen and inoculated intravenously. There is no bioprocess available for production of live attenuated vaccine. The main disadvantage of attenuated vaccines is the possible reversion to virulence, hampering their use in America. A better knowledge of the mechanisms of attenuation of ER is essential to evaluate the risk of using attenuated vaccines on the American mainland.

Inactivated vaccine

This vaccine is based on purified ER elementary bodies which are chemically inactivated and emulsified in an oil adjuvant. Under experimental conditions, this vaccine protects animals against death (with up to 100% survival rates) but does not prevent the disease, since the vaccinated animals exhibit heartwater symptoms such as hyperthermia post-challenge. During field trials, survival rates can vary; and this might be related to the diversity of strains within a restricted area [9]. Nevertheless, vaccine efficacy can be improved by formulating a "regional vaccine", as shown during a field trial in Burkina Faso [9]. The evaluated minimal vaccine efficient dose (35μg) associated with the availability of a semi-industrial process for mass production of the antigen results in a low cost of production (0.14 dollar per dose) [15-17]. A "ready-to-use vaccine" has recently been tested and gives satisfactory results in terms of storage (refrigerated), stability and efficacy (unpublished results).

Identification of suitable vaccines for the American mainland and further needs

A comparison between different types of vaccines is presented in Table 2. This analysis is based on critical criteria: efficacy against homologous/heterologous strains, safety, induction of low morbidity, availability of industrial process, stability, supply, ease of administration and ability to upgrade to a regional vaccine. At the moment, the inactivated vaccine seems to be the most appropriate for the American mainland for four reasons: (1) a bioprocess has already been optimized; (2) killed bacteria are used; (3) storage requirements are compatible with field use; and (4) this vaccine is suitable for upgrades with a regional cocktail of strains to improve its efficacy. As only one Caribbean strain is currently available in culture, it is necessary to isolate and cultivate new Caribbean strains. It will also be essential to characterize and study the genetic and antigenic features of these new strains and compare them to African strains.

Table 2: Analysis of vaccine suitability for the American mainland

Critical Criteria	**Infection & treatment**	**Recombinant vaccine**	**Attenuated vaccine**	**Inactivated vaccine**	**Research needs**
Efficacy/homologous challenge	10	10	10	10	
Efficacy/heterologous challenge	5	**2**	5	5	Genetic & phenotypic characterisation
Safety	**0**	8	**3**	8	Knowledge of virulence mechanisms*
Low morbidity	**0**	8	8	3	
Time to availability	7	7	4	6	
One dose	10	**2**	8	4	
Industrial process	**0**	**0**	**0**	10	Development of bioprocess*
Stability/Shelf Life	8	NT	8	7	
Storage Distribution Supply	**0**	**3**	**0**	8	
Administration	**3**	**2**	**3**	8	
Regional vaccine	**0**	6	**0**	6	Development of regional vaccine: isolation of live Caribbean strains
Total Score	43	48	49	**75**	

NT = not tested

*only for attenuated vaccine

Scoring from 0 to 10, where 0 = the worst and 10 = the best

The genetic and antigenic diversity of ER strains is indeed the main problem for all candidate vaccines mentioned above. The genetic characterization of ER strains circulating in the field in African and Caribbean areas is currently being investigated [9, 18]; and understanding this genetic characterization of ER is essential to clarify the structure of the ER population. However, further improvements are required. Cross-protection studies were performed at the OIE reference laboratory for heartwater (unpublished results) in an attempt to link genotyping to cross-protection, but the correlation has proven to be somewhat difficult to establish. Still, these epidemiological molecular studies will help to select optimal vaccine strains as it has been previously shown [9].

Despite the efforts to develop an effective recombinant vaccine, few genes and/or recombinant proteins have been tested so far. This might be due to the difficulty of selecting protective ER genes, as until now knowledge on ER biology, including virulence mechanisms, is limited. To overcome this issue, virulent and attenuated ER strains from geographically distinct regions are being compared using high-throughput "omic approaches" such as genomics, transcriptomics and proteomics. For example, an ER microarray was recently developed and used to identify the genes differentially expressed during ER development inside host cells [19]; and the first proteomics map of ER elementary bodies (Gardel strain), used in the inactivated vaccine, was also obtained [20].

Conclusions

In the case of heartwater, the most available and widely validated molecular diagnostic assay is *pCS20* nested PCR. Validation of *pCS20* real-time PCR in comparison with nested PCR on multiple strains is now necessary in order to shift from one assay to the other, thereby providing a quicker assay with a significantly reduced risk of cross-contamination. Future emphasis should be given to the development of multi-pathogen molecular diagnostics including different tick-borne pathogens using real-time PCR or microarrays technology.

The inactivated vaccine is currently the most suitable vaccine against heartwater disease in America, especially since the validation of a "ready-to-use" vaccine. In order to develop a vaccine cocktail based on regional strains, new live strains from the Caribbean should be isolated. Due to the high genetic diversity of strains, further genetic characterization using new multi-locus typing must be strengthened both at global and regional levels.

References

1 Dumler J, Barbet A, Bekker C, Dasch G, Palmer G, et al: Reorganization of genera in the families *Rickettsiaceae* and *Anaplasmataceae* in the order *Rickettsiales*: unification of some species of *Ehrlichia* with *Anaplasma*, *Cowdria* with *Ehrlichia* and *Ehrlichia* with *Neorickettsia*, descriptions of six new species combinations and designation of *Ehrlichia equi* and 'HGE agent' as subjective synonyms of *Ehrlichia phagocytophila*. Int J Syst Evol Microbiol 2001;51:2145-2165.

2 Molia S, Frebling M, Vachiery N, Pinarello V, Petitclerc M, et al: *Amblyomma variegatum* in cattle in Marie Galante, French Antilles: prevalence, control measures, and infection by *Ehrlichia ruminantium*. Vet Parasitol 2008;153:338-346.

3 Vachiery N, Jeffery H, Pegram R, Aprelon R, Pinarello V, et al: *Amblyomma variegatum* ticks and heartwater on three Caribbean Islands. Ann NY Acad Sci 2008;1149:191-195.

4 Kasari TR, Miller RS, James AM & Freier JE: Recognition of the threat of *Ehrlichia ruminantium* infection in domestic and wild ruminants in the continental United States. J Am Vet Med Assoc 2010;237:520-530.

5 Porphyre T, Lancelot R, Pegram R, Stachurski F, Lesnoff M, et al: How do temperature and humidity affect ticks population dynamics? A global sensitivity analysis of the tick *Amblyomma variegatum* climatic niche. Ecol Modell:2013; in press.

6 van Vliet AH, van der Zeijst BA, Camus E, Mahan SM, Martinez D, Jongejan F: Use of a specific immunogenic region on the *Cowdria ruminantium* MAP1 protein in a serological assay. J Clin Microbiol 1995;33:2405-2410.

7 Katz JB, DeWald R, Dawson JE, Camus E, Martinez D & Mondry R: Development and evaluation of a recombinant antigen, monoclonal antibody-based competitive ELISA for heartwater serodiagnosis. J Vet Diagn Invest 1997;9:130-135.

8 Martinez D, Vachiery N, Stachurski F, Kandassamy Y, Raliniaina M, et al: Nested PCR for detection and genotyping of *Ehrlichia ruminantium*: use in genetic diversity analysis. Ann NY Acad Sci 2004;1026:106-113.

9 Adakal H, Stachurski F, Konkobo M, Zoungrana S, Meyer DF, et al: Efficiency of inactivated vaccines against heartwater in Burkina Faso: impact of *Ehrlichia ruminantium* genetic diversity. Vaccine 2010;28:4573-4580.

10 Peixoto CC, Marcelino I, Vachiery N, Bensaid A, Martinez D, et al: Quantification of Ehrlichia *ruminantium* by real time PCR. Vet Microbiol 2005;107:273-278.

11 Steyn HC, Pretorius A, McCrindle CM, Steinmann CM, Van Kleef M: A quantitative real-time PCR assay for *Ehrlichia ruminantium* using *pCS20*. Vet Microbiol 2008;131:258-265.

12 Postigo M, Bell-Sakyi L, Paxton E & Sumption K: Kinetics of experimental infection of sheep with *Ehrlichia ruminantium* cultivated in tick and mammalian cell lines. Exp Appl Acarol 2002;28:187-193.

13 Bekker CP, de Vos S, Taoufik A, Sparagano OA, Jongejan F: Simultaneous detection of *Anaplasma* and *Ehrlichia* species in ruminants and detection of E*hrlichia ruminantium* in *Amblyomma variegatum* ticks by reverse line blot hybridization. Vet Microbiol 2002;89:223-238.

14 Allsopp BA: Trends in the control of heartwater. Onderstepoort J Vet Res 2009;76:81-88.

15 Marcelino I, Vachiery N, Amaral AI, Roldao A, Lefrancois T, et al: Effect of the purification process and the storage conditions on the efficacy of an inactivated vaccine against heartwater. Vaccine 2007;25:4903-4913.

16 Marcelino I, Sousa MF, Verissimo C, Cunha AE, Carrondo MJ, Alves PM: Process development for the mass production of *Ehrlichia ruminantium*. Vaccine 2006;24:1716-1725.

17 Vachiery N, Lefrancois T, Esteves I, Molia S, Sheikboudou C, et al: Optimisation of the inactivated vaccine dose against heartwater and *in vitro* quantification of *Ehrlichia ruminantium* challenge material. Vaccine 2006;24:4747-4756.

18 Pilet H, Vachiery N, Berrich M, Bouchouicha R, Durand B, et al: A new typing technique for the *Rickettsiales Ehrlichia ruminantium*: multiple-locus variable number tandem repeat analysis. J Microbiol Methods 2012;88:205-211.

19 Pruneau L, Emboule L, Gely P, Marcelino I, Mari B, et al: Global gene expression profiling of *Ehrlichia ruminantium* at different stages of development. FEMS Immunol Med Microbiol 2012;64(1):66-73.

20 Marcelino I, de Almeida AM, Brito C, Meyer DF, Barreto M, et al: Proteomic analyses of *Ehrlichia ruminantium* highlight differential expression of MAP1-family proteins. Vet Microbiol 2012;156:305-314.

Nathalie Vachiéry, CIRAD, UMR CMAEE, F- 97170 Petit-Bourg, Guadeloupe, France
Tel: +590-590255442, fax: +590-590940396
Email: nathalie.vachiery@cirad.fr

Roth JA, Richt JA, Morozov IA (eds): Vaccines and Diagnostics for Transboundary Animal Diseases. Dev Biol (Basel). Basel, Karger, 2013, vol 135, pp 201-209.

Ebola: Facing a New Transboundary Animal Disease?

F. Feldmann[1], *H. Feldmann*[2]

[1]Office of Operations Management and [2]Laboratory of Virology, Division of Intramural Research, National Institute of Allergy and Infectious Diseases, National Institutes of Health, Rocky Mountain Laboratories, Hamilton, Montana, USA

Key words: Ebola, hemorrhagic fever, pigs, livestock, diagnostics, vaccines

Abstract: Ebola viruses are zoonotic pathogens with the potential of causing severe viral hemorrhagic fever in humans and nonhuman primates. Bats have been identified as a reservoir for Ebola viruses but it remains unclear if transmission to an end host involves intermediate hosts. Recently, one of the Ebola species has been found in Philippine pigs raising concerns regarding animal health and food safety. Diagnostics have so far focused on human application, but enhanced pig surveillance and diagnostics, particularly in Asia, for Ebola virus infections seem to be needed to establish reasonable guidelines for public and animal health and food safety. Livestock vaccination against Ebola seems currently not justified but proper preparedness may include experimental vaccine approaches.

Introduction

Ebola viruses (EBOV) are members of the genus *Ebolavirus* in the family *Filoviridae* of the order *Mononegavirales*. The genus is subdivided into five species, *Zaire ebolavirus*, *Sudan ebolavirus*, *Tai forest ebolavirus*, *Reston ebolavirus*, and the recently discovered tentative species *Bundibugyo ebolavirus* [1]. With the exception of *Reston ebolavirus*, which originates from the Philippines and has not been associated with human disease, all other four species are human pathogens found across Central Africa in a region of approximately 10° north and south of the equator [2].

EBOV causes a disease designated Ebola hemorrhagic fever (EHF) and most outbreaks have been reported from Sudan (*Sudan ebolavirus*), Uganda (*Sudan ebolavirus, Bundibugyo ebolavirus*), Democratic Republic of the Congo (*Zaire ebolavirus, Bundibugyo ebolavirus*), Republic of the Congo (*Zaire ebolavirus*) and Gabon (*Zaire ebolavirus*) [2]. *Tai Forest ebolavirus* has so far caused only one reported severe human infection in West Africa and was retrospectively diagnosed in a survivor in the same region. This leaves three species as the main cause of human EHF, *Zaire*, *Sudan* and *Bundibugyo ebolavirus*. Case fatality rates range between 35-90% based on the species, with *Zaire ebolavirus* being the most virulent one followed by *Sudan* and *Bundibugyo ebolavirus*. Therefore, diagnostic and countermeasure development (antivirals, therapeutics, vaccines) are largely focused on these three EBOV species [2].

EHF is a systemic infection starting with a flu-like disease and often gastrointestinal symptoms. The incubation period ranges from a few days to three weeks with an average of about a week. Infected individuals develop more serious clinical symptoms by the end of the first week after disease onset, which can include multiple forms of hemorrhages accumulating in multiorgan failure and shock. Hemorrhages are not always obvious but their appearance is associated with a bad prognosis. Convalescence can be long with survivors suffering from multiple sequelae, which are not well studied [2, 3].

Transmission among humans is associated with close contact to infected individuals and deceased people during burial procedures. Nosocomial infections occur often due to lack of proper hygiene and safety precautions in hospitals and health care centers in endemic regions [2].

EHF is a zoonosis and certain fruit bat species are thought to be the reservoir [4]. It remains unknown if there are additional reservoirs or a yet unidentified primary reservoir. End hosts for EBOV are largely humans and nonhuman primates with a similar disease outcome. Therefore, hunting and bush meat preparation/consumption are other risk factors for exposure to EBOV [2]. It remains unclear if intermediate or amplifying hosts play a role in the transmission cycle.

Ebola and Livestock

Until very recently, EBOV was only described as a pathogen of humans and nonhuman primates. However, in 2008 *Reston ebolavirus* was discovered in the Philippine pig population causing concerns of food safety and fears of human exposure through livestock [5]. In the meantime a report from China seems to confirm the existence of *Reston ebolavirus* in pigs and expands the region of potential endemicity in Asia [6]. As mentioned before, *Reston ebolavirus* has not yet caused symptomatic infections in humans [2], but infections have been documented in a few individuals involved in handling of nonhuman primates or pigs. Laboratory infection of pigs with *Reston ebolavirus* led to virus replication and shedding from mucosal membranes, but did not cause symptomatic disease in the animals [7]. On the contrary, experimental infection of pigs with *Zaire ebolavirus* resulted in an age-dependent severe disease in pigs with a fulminant lung involvement [8] indicating that EBOV can be pathogenic for this livestock species. Currently, the role of pigs or other livestock species in the transmission of EBOV is unknown, but a matter of great concern to public and animal health. Current *Reston ebolavirus* isolates may be non-pathogenic to humans and pigs, but their close genetic relationship to highly human pathogenic EBOV species and the mutation frequency of an RNA virus, such as EBOV, may well generate and select mutants of a higher virulence for mammals. However, the food safety concern has not been well addressed and urgently needs further attention, particularly if *Reston ebolavirus* is distributed more widely in pig populations throughout Asia.

Diagnostics

The difficulty with EBOV diagnostics is the identification of individual/index cases. The first clinical presentation of disease is rather unspecific and mimics multiple infectious disease causes in tropical Africa as well as developed countries. This calls for a large list of differential diagnostic pathogens and without exposure history to bats, certain wildlife or a known case, single or initial cases are often missed. Therefore, the foremost priority is education of medical personnel to consider EHF in the differential diagnosis and to initiate proper laboratory testing. If case history and clinical presentation are coherent with EHF, proper case patient management including patient isolation may have to be initiated prior to laboratory confirmation [9].

Laboratory diagnostic testing is largely unavailable in endemic and non-endemic areas, but a limited number of reference centers do exist in Africa and worldwide. The use of reference centers delays laboratory confirmation and adds the complication of specimen transport, but in general the diagnostic service is more reliable. Laboratory testing largely follows standard viral diagnostic procedures and is targeted towards detection of the pathogen and the host immune response (see Table 1). Nowadays, molecular detection of viral genome is the assay of choice and multiple systems and approaches are available ranging from RT-PCR/nested PCR to real-time RT-PCR to Loop-Mediated Isothermal PCR (LAMP) and to microarray detection methods. The assays are rapid and highly sensitive, but sometimes prone to failure when it comes to virus variants or new species due to high specificity of primers and probes. The proper specimen sources for molecular detection are whole blood and tissue. Alternatively, swabs derived from mucosal membranes can be used if taken during the symptomatic phase or from deceased people. Antigen detection, usually performed on an enzyme-linked immunosorbent assay (ELISA) platform, is often used in combination with molecular detection assays and serves confirmatory purposes; specimen sources are the same as for molecular detection methods. Detection of the host immune response focuses on antibodies. The assays of choice are IgM-capture ELISA and IgM or IgG direct ELISA. The presence of

Table 1: Ebola Diagnostics

Test System	Target	Specimen (preferred)	Comments
Reverse Transcriptase-Polymerase Chain Reaction (RT-PCR)*	viral genome	blood** / tissue	primary assay
Loop-Mediated Isothermal PCR (LAMP)	viral genome	blood** / tissue	primary assay
Microarray	viral genome	blood** / tissue	alternative assay
In situ-hybridization	viral genome	tissue	alternative assay
Antigen ELISA	viral protein	blood** / tissue	primary assay / confirmatory assay
Fluorescence assay (FA)	viral protein	tissue	alternative assay
Immunohistochemistry (IHC)	viral protein	tissue	alternative assay
Electron microscopy (EM)	viral particle	blood** / tissue	alternative assay
Virus isolation	viral particle	blood** / tissue	alternative assay / confirmatory assay
IgM-capture ELISA	virus-specific antibody	serum	primary assay
IgM direct ELISA	virus-specific antibody	serum	alternative assay
IgG direct ELISA	virus-specific antibody	serum	primary assay
Neutralization assay	virus-specific antibody	serum	alternative assay / confirmatory assay
Immunofluorescence assay (IFA)	virus-specific antibody	serum	alternative assay
Immunoblot	virus-specific antibody	serum	alternative assay

*multiple assays are available and evaluated ranging from RT-PCR/nested PCR to real-time RT-PCR; **whole blood, serum or plasma, (whole blood preferred); ELISA = enzyme-linked immunosorbent assay; primary assay = first choice for diagnostics; alternative assay = not commonly used as a diagnostic assay; confirmatory = used for confirmatory purposes.

EBOV-specific IgM antibodies and a four-fold rise of EBOV-specific IgG antibodies are in general taken as confirmation for a recent infection [9, 10].

All these assays can be performed safely in most diagnostic laboratories on inactivated specimens; the use of high containment (biocontainment level 4 -BSL4) is not necessary for these procedures. Proper inactivation of specimens can be achieved through treatment with chaotropic buffers, irradiation, and a combination of heat/detergent. Virus isolation should certainly be attempted on single or index cases. This, however, requires BSL4 containment and can only be performed in a few facilities worldwide [9, 10]. For outbreak support, mobile laboratories have been deployed that mainly operate on molecular detection technologies and serology but also perform antigen detection ELISA [11].

Current issues with laboratory diagnostics for EBOV are largely related to the unavailability of proper assay systems and control reagents in regular microbiology diagnostic laboratories in endemic areas and worldwide. The easiest and safest methodology is molecular detection and could be established in almost any diagnostic setting. Problems arise with assay evaluation, quality assurance and test performance. For single/index cases it is therefore highly recommended to seek independent confirmation by a reference laboratory, given the tremendous impact of a false-positive or false-negative result. It should be mentioned that with the exception of one assay (Ebola abicap antigen detection; Senova), no other assays are commercially available.

Vaccines

The development of countermeasures against pathogens of bioterrorism, of which EBOV is a prime example, has dramatically improved over the past decade due to increased interest and funding by governments of developed countries. This work is highly dependent on animal infection/disease models which are important components of the licensing procedure for modalities against pathogens for which human efficacy trials cannot be performed. Multiple animal models have been developed for EBOV [12, 13]. The rodent models (mouse, guinea pig and hamster) are dependent on in vivo adaptation. Adapted strains usually show mutations in proteins that interfere with the interferon response of the host, such as the EBOV VP30 and VP24 proteins [2, 13]. The rodent models are disease models but lack certain key characteristic features of human EHF. Among the rodent models, the mouse is a good screening model due to ease of handling and the availability of immunological tools. The predictive value for efficacy testing, however, is low which might be explained by a lack of several hallmark features of human EHF. The guinea pig model has a higher predictive value for efficacy testing of countermeasures likely due to more similar clinical features with human EHF, but it is limited in its application by the general lack of tools available for this animal species. The recently developed hamster model, for which tools are currently under development, shows great potential, because it seems to mimic human EHF best of all rodent models, including the development of coagulation abnormalities that are not observed in mice and guinea pigs. This model, however, has not yet been used for extensive efficacy testing of countermeasures. The ultimate animal model is the nonhuman primate, in particular cynomolgus and rhesus macaques, which show all hallmarks of human EHF and are considered to have the best predictive value for efficacy in humans due to their close genetic relationship.

Efforts to develop EBOV vaccines go back to the early '80s of the last century when killed virus vaccines were produced, however, with low efficacy in protecting nonhuman primates [2, 14, 15] (see Table 2). This was followed by DNA vaccination and the use of subunit vaccines based on either purified EBOV proteins or vector-expressed EBOV proteins (i.e. vaccinia virus, baculovirus), mainly the sole surface glycoprotein (GP), which is the target for neutralizing antibodies. Despite promising results with some approaches in rodent models, efficacy testing in nonhuman primates largely failed. Funding of countermeasures against biothreat pathogens has changed the concept of vaccine development in the field of rare emerging/re-emerging pathogens. Emergency vaccines are of higher priority with features such as single vaccination, short-time to immunity, long-lasting immunity and ease of application being important criteria. More recent approaches include attenuated replication-deficient and -competent viral vectors expressing GP as the sole immunogen or in combination with certain other viral structural proteins. Today, the leading platforms with good efficacy in nonhuman primate models are replication-deficient adenovirus (Ad5) and replication-competent vesicular stomatitis virus (VSV) vectors expressing the EBOV GP (different vectors for the distinct EBOV species). These vectors have resulted in 100% protection of macaques against lethal homologous (same species) challenge following a single-shot immunization approach. Unfortunately, cross-species protective efficacy is rather low, but can be overcome by blending of vectors expressing GPs from different EBOV species. In case of VSV, the vaccine has also shown complete or partial protective efficacy (species-dependent) when used up to 48 hours after lethal EBOV challenge [16]. In addition, virus-like particles (VLPs), minimally consisting of the EBOV matrix protein VP40 and GP, have shown promising efficacy in macaque models of EHF [17]. The biggest advantage of VLPs over the most promising vector-based platforms is their safety record (potential problem with VSV-based vaccines) and the lack of pre-existing immunity (problem with Ad5-based vaccines), but in contrast to the vector-based platforms, VLP-based vaccines require booster immunization, which limits their use in emergency situations. Currently, only the adenovirus-based vectors and DNA-based vaccines are in clinical trials.

Conclusions

Given the more recent development with evidence of *Reston ebolavirus* in pigs in certain Asian countries, EBOV may be considered a potential livestock pathogen

Table 2: Ebola Vaccines

Vaccine Platform	Characteristic	Efficacy	Status
Adenovirus-based (Ad5 & other serotypes) (GP)	replication-deficient	nonhuman primates (pre-exposure; single shot)	clinical trial
Vesicular stomatitis virus (VSV)-based (GP)	replication-competent	nonhuman primates (pre- & post-exposure; single shot)	experimental
Virus-like particle (GP & VP40 & NP)	replication-incompetent	nonhuman primates (pre-exposure; boosts)	experimental
Paramyxovirus-based (HPIV) (GP)	replication-competent	nonhuman primates (pre-exposure; single shot)	experimental
Virus-like replicon particle (VRP) (GP)	replication-deficient	nonhuman primates (pre-exposure; boost)	experimental
Subunit (GP)	replication-incompetent	rodents (pre-exposure; boost)	experimental
DNA (GP & N)	replication-incompetent	rodents (pre-exposure; boost)	clinical trial

Ad5 = adenovirus serotype 5; GP = Ebola glycoprotein; HPIV = human parainfluenzavirus; N = Ebola nucleoprotein; VP40 = Ebola matrix protein; VSV = vesicular stomatitis virus, serotype Indiana

with transboundary threat potential. Testing for *Reston ebolavirus* or EBOV-like viruses in pigs may be considered for certain countries as well as for countries that import from areas that have reported infections. In general, it should not be a problem to adapt the existing diagnostic tests for use in livestock species such as pigs, but a commercial test would certainly improve the use in the veterinary field and provide better quality diagnostic results.

In order to provide better guidance for authorities in public and animal health we need more surveillance data on pig populations in Asia and other parts of the world.

Further experimental studies have to address the virulence of *Reston ebolavirus* or other EBOVs in pigs and perhaps other livestock species. We also need to study the transmission potential among pigs and the potential of cross-species transmission into other mammalian species including humans. Interestingly, transmission from pigs to nonhuman primates has recently been observed during experimental infection of pigs with *Zaire ebolavirus* [18].

Vaccination of pigs against EBOV, however, seems premature at this time and currently not justified. Nevertheless, experimental vaccine approaches could be developed and tested in pig models as a measure for preparedness, should the situation change with more evidence for infection in expanded pig populations. The current vaccine approaches developed for human applications may not all be best suited for livestock applications and new approaches should be evaluated using vaccine vector systems with good safety and efficacy profiles in livestock species.

Overall, the potential threat posed to livestock and humans through EBOV or EBOV-like viruses in pigs should not be underestimated and certainly needs further attention and careful follow-up.

Acknowledgements

The work on Ebola viruses at the Integrated Research Facility of the Rocky Mountain Laboratories in Hamilton, Montana is funded by the Intramural Research Program of the National Institute of Allergy and Infectious Diseases (NIAID) of the National Institutes of Health (NIH). Opinions expressed in this paper are those of the authors and not of the NIAID, NIH.

References

1 Kuhn JH, Becker S, Ebihara H, Geisbert TW, Jahrling PB, Kawaoka Y, et al: Filoviridae, in: King AMQ, Adams MJ, Carstens EB, Lefkowitz EJ (eds): Virus Taxonomy – Ninth Report of the International Committee on Taxonomy of Viruses, Elsevier Academic Press, 2012, 9th edition, pp 665-671.

2 Feldmann H, Geisbert TW: Ebola haemorrhagic fever. Lancet 2011;377:849-862.

3 Pattyn SR (ed): Ebola Virus Haemorrhagic Fever, Elsevier, Amsterdam North-Holland, 1978, vol 1, pp 1–280.

4 Leroy EM, Kumulungui B, Pourrut X, Rouquet P, Hassanin A, Yaba P, et al: Fruit bats as reservoirs of Ebola virus. Nature 2005;438:575-576.

5 Barrette RW, Metwally SA, Rowland JM, Xu L, Zaki SR, Nichol ST, et al: Discovery of swine as a host for the Reston ebolavirus. Science 2009; 325:204-206.

6 Pan Y, Zhang W, Cui L, Hua X, Wang M, Zeng Q: Reston virus in domestic pigs in China. Arch Virol 2012;DOI 10.1007/s00705-012-1477-6.

7 Marsh GA, Haining J, Robinson R, Foord A, Yamada M, Barr JA, et al: Ebola Reston virus infection of pigs: clinical significance and transmission potential. J Infect Dis 2011;204 (Suppl 3):S804-S809.

8 Kobinger GP, Leung A, Neufeld J, Richardson JS, Falzarano D, Smith G, et al: Replication, pathogenicity, shedding, and transmission of Zaire ebolavirus in pigs. J Infect Dis 2011;204:200-208.

9 Sanchez A, Geisbert TW, Feldmann H: Filoviridae: Marburg and Ebola viruses, in Knipe DM, Howley PM (eds): Fields Virology. Lippincott Williams & Wilkins, Philadelphia, 2006, pp1409-1448.

10 Strong JE, Grolla A, Jahrling PB, Feldmann H: Filoviruses and arenaviruses, in Detrick B, Hamilton RG, Folds JD (eds): Manual of Molecular and Clinical Laboratory Immunology, ASM Press, 2006, 7th edition, pp774-790.

11 Grolla A, Jones SM, Fernando L, Strong JE, Ströher U, Möller P, et al: The use of a mobile laboratory unit in support of patient management and epidemiological surveillance during the 2005 Marburg Outbreak in Angola. PLoS Negl Trop Dis 2011;5:e1183.

12 Bente D, Gren J, Strong JE, Feldmann H: Disease modeling for Ebola and Marburg viruses. Dis Model Mech 2009;2:12-17.

13 Ebihara H, Zivcec M, Gardner D, Falzarano D, Lacasse R, Rosenke R, et al: A Syrian golden hamster model recapitulating Ebola hemorrhagic fever. J Infect Dis 2012 Oct 8. [Epub ahead of print]

14 Fausther-Bovendo H, Mulangu S, Sullivan NJ: Ebolavirus vaccines for humans and apes. Curr Opin Virol 2012;2:324-329.

15 Hoenen T, Groseth A, Feldmann H: Current Ebola vaccines. Expert Opin Biol Ther 2012;12:859-872.

16 Geisbert TW, Feldmann H: Recombinant vesicular stomatitis virus-based vaccines against Ebola and Marburg virus infections. J Infect Dis 2011;204 (Suppl 3):S1075-S1081.

17 Warfield KL, Aman MJ: Advances in virus-like particle vaccines for filoviruses. J Infect Dis 2011;204 (Suppl 3):S1053-S1059.

18 Weingartl HM, Embury-Hyatt C, Nfon C, Leung A, Smith G, Kobinger G. Transmission of Ebola virus from pigs to non-human primates. Sci Rep. 2012;2:811. doi: 10.1038/srep00811.

Heinz Feldmann, MD, Laboratory of Virology, Division of Intramural Research (DIR), National Institute of Allergy and Infectious Diseases (NIAID), National Institutes of Health (NIH), Rocky Mountain Laboratories, Hamilton, Montana, USA

Tel.: +1-406-375-7410, fax: +1-406-375-7416

Email: feldmannh@niaid.nih.gov

Roth JA, Richt JA, Morozov IA (eds): Vaccines and Diagnostics for Transboundary Animal Diseases. Dev Biol (Basel). Basel, Karger, 2013, vol 135, pp 211-218.

Review of Ebola Virus Infections in Domestic Animals

H.M. Weingartl[1]*, *C. Nfon*[1], *G. Kobinger*[2]

[1] National Centre for Foreign Animal Disease (NCFAD), Canadian Food Inspection Agency (CFIA), Canadian Science Centre for Human and Animal Health, Winnipeg, Canada

[2] National Microbiology Laboratory (NML), Public Health Agency of Canada (PHAC), Canadian Science Centre for Human and Animal Health, Winnipeg, Canada

Key words: Ebola virus, dog, pig, pathogenesis, diagnostic test

Abstract: Ebola viruses (EBOV; genus *Ebolavirus*, family *Filoviridae*) cause often fatal, hemorrhagic fever in several species of simian primates including human. While fruit bats are considered a natural reservoir, the involvement of other species in the EBOV transmission cycle is unclear, especially for domesticated animals. Dogs and pigs are so far the only domestic animals identified as species that can be infected with EBOV. In 2009 Reston-EBOV was the first EBOV reported to infect swine with indicated transmission to humans; and a survey in Gabon found over 30% seroprevalence for EBOV in dogs during the Ebola outbreak in 2001-2002. While infections in dogs appear to be asymptomatic, pigs experimentally infected with EBOV can develop clinical disease, depending on the virus species and possibly the age of the infected animals. In the experimental settings, pigs can transmit Zaire-Ebola virus to naive pigs and macaques; however, their role during Ebola outbreaks in Africa needs to be clarified. Attempts at virus and antibody detection require as a prerequisite validation of viral RNA and antibody detection methods especially for pigs, as well as the development of a sampling strategy. Significant issues about disease development remain to be resolved for EBOV. Evaluation of current human vaccine candidates or development of veterinary vaccines de novo for EBOV might need to be considered, especially if pigs or dogs are implicated in the transmission of an African species of EBOV to humans.

Introduction

Ebola viruses (EBOV) are enveloped, single-stranded, negative sense RNA viruses classified into the family *Filoviridae*. The African species of EBOV, *Zaire*, *Sudan* and *Bundibugyo ebolaviruses* cause severe haemorrhagic fevers in humans, with high fatality rates. *Cote d'Ivoire ebolavirus* is considered to be less virulent in humans, while *Reston ebolavirus*, the only species found outside Africa, has never been associated with human disease, despite documented exposures [1-3].

Accumulated evidence has confirmed fruit bats as the natural reservoir of EBOV [4, 5], while several wildlife species have been identified as susceptible to infection with EBOV; and bushmeat has been implicated in the initiation of some human outbreaks. However, the role of domestic animals is not yet clear, and to date only limited sampling has been conducted during or shortly after Ebola outbreaks seeking to determine whether domestic animals can be infected with Ebola virus and possibly participate in the virus transmission cycle. From five tested species – bovine, ovine, caprine, swine and canine – only dogs were found to have antibodies against EBOV [6, 7]. Recently, Reston-Ebola virus (REBOV) has been isolated from pigs in the Philippines, and importantly with further transmission to humans, which raises a concern that this species may have been missed as susceptible to EBOV during the surveys due to the number of samples being limited to only twelve pigs.

Infection of Swine with REBOV

In 2008, REBOV was detected in pigs in the Philippines. The Philippine Department of Agriculture requested assistance from the US Department of Agriculture Foreign Animal Disease Diagnostic Laboratory (FADDL) and submitted samples for laboratory diagnostics, because of the suspected outbreak of highly pathogenic porcine reproductive and respiratory syndrome (PRRS) with high mortality on the affected swine farms. FADDL did indeed confirm infection with PRRSV, with some pigs also being co-infected with porcine circovirus 2. In addition, the laboratory detected an L gene of REBOV on the panviral microarray. Therefore, samples were transferred to CDC Atlanta; and the presence of REBOV was confirmed [8]. Subsequently, antibodies to the virus were found in several pig farmers, providing a strong indication that they had acquired the virus from infected pigs [8, 9].

Experimental inoculation of swine with REBOV did not reproduce the high mortality observed in the Philippines, suggesting that the mortality in swine was likely due to the highly pathogenic PRRSV. Marsh and co-authors infected oronasally eight 5-6 weeks old piglets with 10^6 $TCID_{50}$ of the Philippines 2008 isolate of REBOV, and eight piglets of the same age subcutaneously with the same dose of virus [10]. Shedding from the nasopharynx at the RNA and infectious virus levels was detected in all pigs inoculated oronasally. The virus was detected in multiple organs in all inoculated pigs, although the piglets remained clinically healthy [10].

Infection of Swine with ZEBOV

Infections of swine with the African species of *ebolavirus* have not been reported; however, in the experimental infection studies, *Zaire ebolavirus* (ZEBOV) appears to be far more pathogenic than REBOV in the same age group of pigs. Domestic pigs (Landrace breed) challenged by oronasal and ocular routes with a total of $1x10^6$ PFU of ZEBOV developed severe respiratory distress requiring euthanasia for humane reasons. Under similar experimental conditions, virus transmission from infected to naive animals was evaluated in a second set of pigs. These piglets were one week younger than the first group, and interestingly developed only mild clinical signs, but transmitted the virus to co-housed piglets of the same age [11]. The two experiments were repeated, this time inoculating piglets from the same litter – six at the age of 4 weeks and six at the age of 6 weeks. The clinical outcomes were strikingly similar: younger piglets displayed mild to moderate clinical signs, while the older piglets again developed rather severe respiratory distress [12, 13]. Interestingly, the age difference in development of clinical disease may be related to higher levels of interferon-α at the time of inoculation with EBOV in the younger piglets (mean value of 60 pg/ml of serum compared to 30 pg/ml in the older piglets) (unpublished), an observation worth further investigation.

Transmission of ZEBOV from infected piglets was demonstrated not only to all naive pigs co-habiting with inoculated animals [11], but also to all cynomolgus macaques housed in an open inaccessible cage system in the same room as the infected piglets [12].

Pathogenesis of EBOV in Swine

Infection of swine with EBOV is systemic with major involvement of the respiratory tract as well as pathology in the lungs. The major route of virus shedding is from the nasopharynx, especially for infections acquired via the oronasal route. The severity of pathological changes and the development of clinical signs are likely to depend on the infecting species of *ebolavirus*, as well as the age, and possibly also on the genetic background of the pigs.

Following mucosal exposure, pigs replicated ZEBOV to high titers (reaching 10^7 $TCID_{50}$/ml), mainly in the respiratory tract, and also developed severe lung pathology. Oronasal shedding was detected for up to 14 days post-infection (dpi); and the virus was also briefly detected in blood. Neutralizing antibodies were detected only at around 10 dpi, while the first IgM antibody detection by ELISA occurred between 5 and 7 dpi [11,13]. Oronasal shedding was confirmed for REBOV infected piglets, but the lung pathology was considerably less severe; and the virus was not detected in the blood. IgG antibody development was detected using ELISA in subcutaneously inoculated piglets at 8 dpi, when only 2 out of 4 oronasally infected pigs seroconverted. All pigs were considered positive for EBOV IgG antibody at 10 dpi [10], similar to the antibody development against ZEBOV.

In addition to lung and nasal turbinates, the infectious virus or viral RNA of either ZEBOV or REBOV was detected in the liver, spleen, tonsil, bronchiolar and submandibular lymph nodes, and at low levels in the heart, kidney and intestine. REBOV RNA was also detected in skeletal muscle, but the RNA detection attempts in the muscle of ZEBOV infected piglets were negative [10, 11].

Although alveolar macrophages appear to be the most important target for EBOV in the lungs of both piglets and non-human primates, the observed outcome of the infection of macrophages in terms of histopathological lesions in the lungs is notably different. Almost no infiltration of other immune cells into the lungs was observed for macaques infected with ZEBOV [12] or pigs infected with REBOV [10], while massive infiltration of not only macrophages but also neutrophils and lymphocytes was observed in the lungs of older piglets infected with ZEBOV, although antigen was present only in the alveolar macrophages. The infiltration had essentially reached the point when no alveolar space was left, and lead to necrosis 5-7 days post-infection [11]. Microarray analysis of changes in gene expression in lung tissue from the older ZEBOV-infected piglets detected upregulation of a high number of proinflammatory cytokines, chemokines and acute phase protein genes, known to attract immune cells to sites of infection, essentially resulting in a cytokine storm [13]. The detected upregulation of the anti-inflammatory cytokine IL-10 genes in ZEBOV-infected lungs did not appear to be sufficient to control the inflammatory process. An overactivation of the pulmonary proinflammatory response likely plays a crucial role in the pathogenesis of ZEBOV infection in pigs through an out of control attraction of inflammatory cells to the lungs in the experimentally inoculated piglets older than 5 weeks.

Systemic effects included a decline in the proportion of monocyte/dendritic and B cells in peripheral blood [13], which further contributed to the disregulation of the adaptive immune response noticeable because of a somewhat delayed production of antibodies, with the IgG class antibodies and the virus neutralizing antibodies detected only about two weeks post-infection [10, 11], as described above.

Diagnostic Assays

A critical aspect of reliable diagnostics is a well-designed approach to sample collection (Fig. 1). Since REBOV could not be detected in the blood of infected pigs, nasopharyngeal swabs collected between 4 - 8 days post infection might be the only live sampling available for virus detection. This leaves a period of about a week when infection with REBOV would not be detected, until the development of antibodies around 2 weeks post-infection [10]. From a diagnostic point of view, the situation with ZEBOV is better than with REBOV, as shedding ceases with the

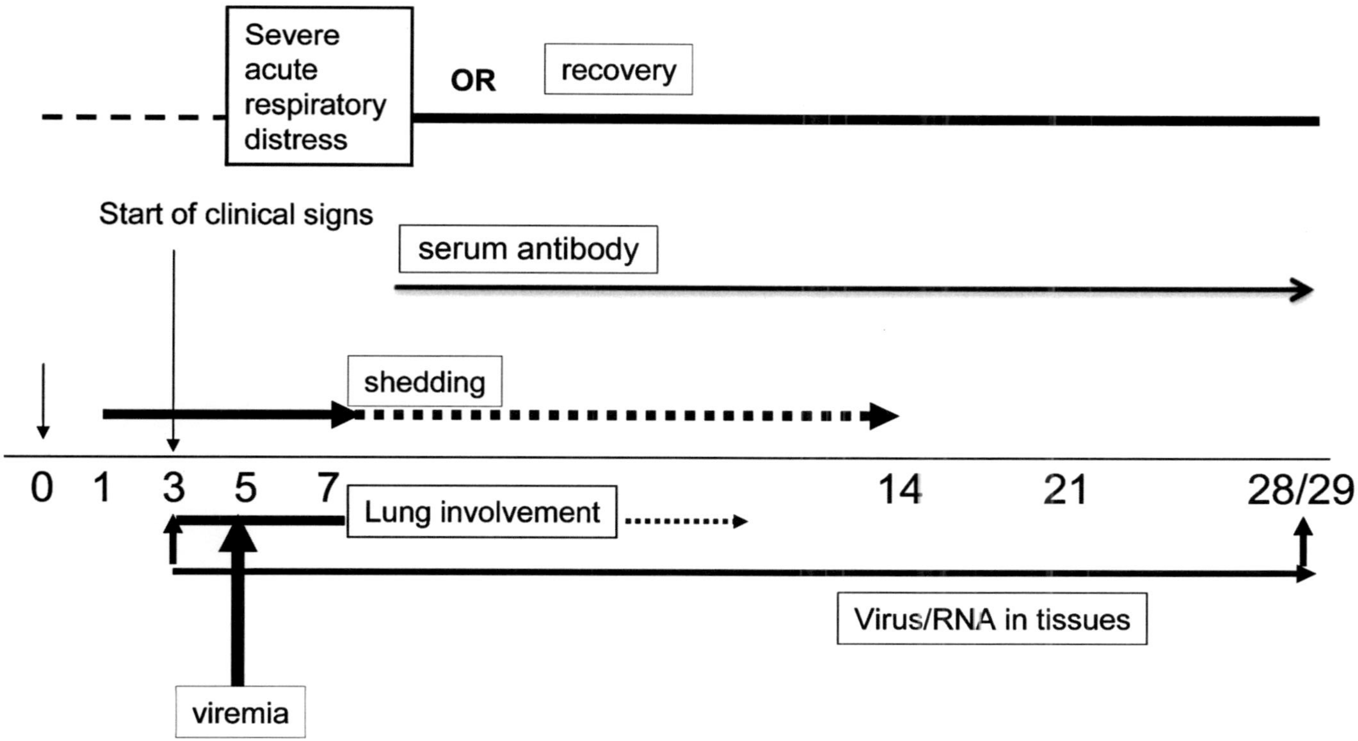

Fig. 1. Overview of current understanding of disease development and sampling strategy for virus detection in ZEBOV infected swine.

development of antibodies; and virus RNA is detectable in the blood, at least briefly, around 5 days post-infection [11, 13]. However, the duration of detectable levels of antibodies in pigs is not known.

Antigen detection

Antigen detection by immunohistochemistry in fixed tissues, especially lungs and lymph nodes using rabbit or mouse polyclonal anti-EBOV antibodies appears to be reliable [8, 10, 11]. An approach using monoclonal antibodies might need to be considered for consistent long-term testing.

Viral RNA detection

Viral RNA detection was performed using different real-time RT-PCR assays targeting the L gene [12], the glycoprotein gene [11] or the nucleoprotein gene [10]. All these approaches were sensitive in swine samples, as well as comparable between laboratories where applicable, e.g. between the National Microbiology Laboratory (NML) and the National Centre for Foreign Animal Disease (NCFAD), with differences laying more in respective biosafety protocols as well as in different instrumentation between the laboratories. Harmonization may be required, when necessary.

Virus isolation

Vero E6 cells are currently the preferred cell substrate for EBOV isolations. Although virus stock preparation did not show any difference between Vero E6 cells from NML compared to NCFAD cells, EBOV isolation from swine swabs and tissues did indicate differences in the formation of the cytopathic effect (CPE) between the two laboratories. EBOV re-isolated from swine did replicate, based on progressively increasing genomic copy number during the passage of the isolates, but did not cause CPE in the NCFAD Vero E6 cells until the second or third passage. Therefore, blind passages during isolation attempts from field samples might be required. Currently, swine kidney cell line (PK15) is under evaluation as a candidate for virus isolation from swine samples. The classical CPE-reduction virus neutralization assays (detection of neutralizing antibodies in serum) in Vero E6 cells are suitable for swine serum samples.

Antibody detection

Antibody detection by ELISA employing cell lysate antigen [14] was successfully used in the serosurvey for EBOV antibodies in dogs [7]. This ELISA format, using CDC antigen, was used in IgM capture assay in pig serum by Nfon et al [13], as well as in the study by Marsh and colleagues [10] to detect IgG antibodies in their experimentally infected piglets. IgG detection using ELISA employing the inactivated purified whole virions as an antigen was also successfully used for swine serum [11]. ELISAs for antibody detection in swine sera will likely require regional validation to determine diagnostic specificity in swine. An understanding of the dynamics of EBOV antibody development in pigs would be important; especially if use of recombinant antigens would be considered in designing ELISA assays for lower containment laboratories.

At the present time, there are many unanswered questions about disease development caused by EBOV in swine. For example: the extent to which the development of different Ebola virus infections is linked to the age of the infected animal or the extent to which the virus can be shed without apparent disease. Despite the growing body of knowledge about the nature of Ebola viruses, it remains uncertain whether the transmission of these viruses can be sustained in a swine herd. Even more disturbing is a further unanswered question: Can the evolution or adaptation of REBOV in swine lead to mutations which would allow for the virus to be pathogenic in humans? In view of the fact that viruses can be present in the semen of other species, the possibility that artificial insemination might lead to extensive EBOV transmission among swine needs to be investigated.

Conclusion

The experimental studies cited above have confirmed that pigs infected with REBOV shed the virus without clinical signs; and this shedding may also apply to some pigs infected with ZEBOV [10, 12]. In addition, clinical signs of Ebola virus infections are not specific and resemble other infections of the respiratory system, e.g. PRRSV. Infected animals pose a transmission risk to humans [10,12]. The Food and Agricultural Organization (FAO) pig density map for Uganda (http://geonetworkfao.org) shows that outbreaks of Ebola virus have taken place in areas of great pig density, raising concern about potential of ZEBOV epidemics due to transmission of the virus to pigs, and subsequently to humans. It is now evident that the infection of an index human case would be sufficient to introduce the Ebola virus into the human population, the evaluation of current human vaccine candidates or the development of veterinary vaccines for EBOV de novo might need to be considered as well as adaptation and evaluation of existing diagnostic tests for veterinary purposes.

Acknowledgements

The animal work for these studies has been done by H. Weingartl, G. Kobinger and K. Tierney, with histopathology by J. Neufeld, C. Embury-Hyatt and E. Moffat, in vitro live virus work by C. Nfon, A. Leung and G. Smith, and further lab work by J. Richardson, D. Falzarano and A. Patel. We are grateful to funding from PHAC and CRTI, project number 3780-2011-3ova-17.

References

1 Hayes CG, Burans JP, Ksiazek TG, Del Rosario RA, Miranda ME, Manoloto CR, et al: Outbreak of fatal illness among captive macaques in the Philippines caused by an Ebola-related filovirus. Am J Trop Med Hyg 1992;46:664-671.

2 Miranda ME, Miranda NL: Reston ebolavirus in humans and animals in the Philippines: a review. J Infect Dis 2011;204(Suppl 3):S757-760.

3 Feldmann H, Geisbert TW: Ebola haemorrhagic fever. Lancet 2011;377(9768): 849-862.

4 Leroy EM, Kumulungui B, Pourrut X, Pourrut X, Rouquet P, Hassanin A, et al: Fruit bats as reservoirs of Ebola virus. Nature 2005;438:575-576.

5 Leroy EM, Epelboin A, Mondonge V, Pourrut X, Gonzalez JP, Muyembe-Tamfum JJ, Formenty P: Human Ebola outbreak resulting from direct exposure to fruit bats in Luebo, Democratic Republic of Congo, 2007. Vector Borne Zoonotic Dis 2009;9:723-728.

6 Olson SH, Reed P, Cameron KN, Ssebide BJ, Johnson CK, Morse SS, et al: Dead or alive: animal sampling during Ebola hemorrhagic fever outbreaks in humans. Emerg Health Threats J 2012;5:doi:10.3402/ehtj.v50i0.9134.

7 Allela L, Bourry O, Pouillot R, Délicat A, Yaba P, Kumulungui B, et al: Ebola virus antibody prevalence in dogs and human risk. Emerg Infect Dis 2005;11:385-390.

8 Barette RW, Metwally SA, Rowland JM, Xu L, Zaki SR, Nichol ST, et al: Discovery of swine as a host for the Reston ebolavirus. Science 2009;325:204-206.

9 WHO: Ebola Reston in pigs and humans, Phillipines. Wkly Epidemiol Rec 2009;84:49-50.

10 Marsh GA, Haining J, Robinson R, Foord A, Yamada M, Barr JA, et al: Ebola Reston virus infection of pigs: clinical significance and transmission potential. J Infect Dis 2011;204:S804-S809.

11 Kobinger GP, Leung A, Neufeld J, Richardson JS, Falzarano D, Smith G, et al: Replication, pathogenicity, shedding, and transmission of Zaire ebolavirus in pigs. J Infect Dis 2011;204:200-208.

12 Weingartl HM, Emburry-Hyatt C, Nfon C, Leung C, Smith G, Kobinger G: Transmission of Ebola virus from pigs to non-human primates. Sci Rep 2012;2:811.

13 Nfon CK, Leung A, Smith G, Embury-Hyatt C, Kobinger G, Weingartl HM: Chemokines and proinflammatory cytokines in the pathogenesis of pulmonary disease in Zaire Ebola virus-infected pigs. PLoS One 2013; in press.

14 Ksiazek TG, West CP, Rollin PE, Jahrling PB, Peters CJ: ELISA for the detection of antibodies to Ebola viruses. J Infect Dis 1999;179:S92-S97.

Hana Weingartl, NCFAD, CFIA, 1015 Arlington St., Winnipeg, MB, R3E 3M4, Canada
Tel.: 204 789 2027, fax: 204 789 2038
Email: hana.weingartl@inspection.gc.ca

Index of Authors

Developments in Biological Standardization

Vol. 36	Washington DC, 1976	Freeze-Drying of Biological Products
Vol. 37	Geneva, 1976	Standardization of Cell Substrates for the Production of Virus Vaccines
Vol. 38	London, 1977	Biological Preparations in the Treatment of Cancer
Vol. 39	Geneva, 1977	Influenza Immunization
Vol. 40	Marburg, 1977	Standardization of Rabies Vaccines for Human Use Produced in Tissue Culture (Rabies III)
Vol. 41	Guadeloupe, 1978	Vaccines in the Developing Countries
Vol. 42	Paris, 1978	Second General Meeting of ESACT
Vol. 43	Brussels, 1978	Immunization: Benefit Versus Risk Factors
Vol. 44	Villars, 1979	Test methods for the Quality Control of Plasma Proteins
Vol. 45	San Antonio, 1979	The Standardization of Animals to Improve Biomedical Research, Production and Control
Vol. 46	Oxford, 1979	Third General Meeting of ESACT
Vol. 47	Bilthoven, 1980	Reassessment of Inactivated Poliomyelitis Vaccine
Vol. 48	Geneva, 1980	Standardization of Albumin, Plasma Substitutes and Plasmapheresis
Vol. 49	Leetown, 1981	Fish Biologics: Serodiagnostics and Vaccines
Vol. 50	Heidelberg, 1981	The Use of Heteroploid and Other Cell Substrates for the Production of Biologicals
Vol. 51	Lyons, 1981	Immunization of Adult Birds with Inactivated Oil Adjuvant Vaccines
Vol. 52	Lyons, 1981	Herpes Virus of Man and Animal: Standardization of Immunological Procedures
Vol. 53	Dublin, 1982	Enteric Infections in Man and Animals: Standardization of Immunological Procedures
Vol. 54	Athens, 1982	Viral Hepatitis: Standardization in Immunoprophylaxis of Infections by Hepatitis Viruses
Vol. 55	Copenhagen, 1982	Fifth General Meeting of ESCAT
Vol. 56	Algiers, 1983	Third International Symposium on Brucellosis
Vol. 57	Paris, 1983	Monoclonal Antibodies: Standardization of their Characterization and Use
Vol. 58	Budapest, 1983	International Symposium on BCG Vaccines and Tuberculins
Vol. 59	Geneva, 1983	Standardization and Control of Biologicals Produced by Recombinant DNA Technology
Vol. 60	Gardone Riviera, 1984	Production and Exploitation of Existing and New Animal Cell Substrates
Vol. 61	Geneva, 1984	Proceedings of the Fourth International Symposium on Pertussis
Vol. 62	Stockholm, 1985	Diagnostics and Vaccines for Parasitic Diseases
Vol. 63	San Francisco, 1985	Use and Standardization of Chemically Defined Antigens
Vol. 64	London, 1985	Reduction of Animal Usage in the Development and Control of Biological Products

Vol. 65	Amsterdam, 1985	Use and Standardization of Combined Vaccines
Vol. 66	Vienna, 1985	Advances in Animal Cell Technology: Cell Engineering, Evaluation and Exploitation
Vol. 67	Melbourne, 1986	Standardization in Blood Fractionation Including Coagulation Factors
Vol. 68	Geneva, 1986	Cells - Products - Safety
Vol. 69	London, 1987	Cytokines Laboratory and Clinical Evaluation
Vol. 70	Arlington, VA, 1988	Continuous Cell Lines as Substrates for Biologicals
Vol. 71	Utrecht, 1989	Monoclonal Antibodies for Therapy, Prevention and In vivo Diagnosis of Human Disease
Vol. 72	Annecy, 1989	Progress in Animal Retroviruses
Vol. 73	Shizuoka, 1990	Pertussis: Evaluation and Research on Acellular Pertussis Vaccines
Vol. 74	Bethesda, MD, 1990	Biological Product Freeze-Drying and Formulation
Vol. 75	London, 1990	Virological Aspects of the Safety of Biological Products
Vol. 76	Bethesda, MD, 1991	Continuous Cell Lines - An International Workshop on Current Issues
Vol. 77	Annecy, 1991	Standardization of the Immunopharmacology of Natural and Synthetic Immunomodulators
Vol. 78	Bethesda, MD, 1991	Poliovirus Attenuation: Molecular Mechanisms and Practical Aspects
Vol. 79	Ploufragan, 1992	The First Steps Towards an International Harmonization of Veterinary Biologicals: 1993 and Free Circulation of Vaccines within the EEC
Vol. 80	Heidelberg, 1992	Transmissible Spongiform Encephalopathies - Impact on Animal and Human Health
Vol. 81	Cannes, 1992	Virological Safety Aspects of Plasma Derivatives
Vol. 82	Albany NY, 1993	Recombinant Vectors in Vaccine Development
Vol. 83	Annecy, 1993	Genetic Stability and Recombinant Product Consistency
Vol. 84	Langen, 1993	Non-Target Effects of Live Vaccines
Vol. 85	Santa Fé, NM, 1994	Genetics of Streptococci, Enterococci and Lactococci
Vol. 86	Langen, 1994	Replacement, Reduction and Refinement of Animal Experiments in the Development and Control of Biological Products
Vol. 87	Geneva, 1995	New Approaches to Stabilisation of Vaccines Potency
Vol. 88	Bethesda, MD, 1995	Viral Safety and Evaluation of Viral Clearance from Biopharmaceutical Products
Vol. 89	Rome, 1995	Pertussis Vaccine Trials
Vol. 90	Oslo, 1996	Fish Vaccinology
Vol. 91	San Francisco, CA, 1996	Development of Specifications for Biotechnology Pharmaceuticals Products
Vol. 92	Bergen, 1996	Modulation of the Immune Response to Vaccine Antigens
Vol. 93	Annecy, 1996	Safety of Biological Products Prepared from Mammalian Cell Culture
Vol. 94	Bethesda, MD, 1997	Simian Virus 40 (SV40): A Possible Human Polyomavirus
Vol. 95	Paris, 1997	Preclinical and Clinical Development of New Vaccines
Vol. 96	Washington, DC, 1995	Characterization of Biotechnology Pharmaceutical Products
Vol. 97	London, 1997	Biological Characterization and Assay of Cytokines and Growth Factors
Vol. 98	London, 1997	Inactivated Influenza Vaccines Prepared in Cell Culture
Vol. 99	Strasbourg, 1998	Animal Sera, Animal Sera Derivatives and Substitutes Used in the Manufacture of Pharmaceuticals: Viral Safety and Regulatory Aspects
Vol. 100	Geneva, 1998	A Celebration of 50 Years of Progress in Biological Standardization and Control at WHO
Vol. 101	London, 1998	Alternatives to Animals in the Development and Control of Biological Products for Human and Veterinary Use

Developments in Biologicals

Vol. 102 San Francisco, CA, 1999	Advances in Transfusion Safety
Vol. 103 Annecy, 1999	Physico-Chemical Procedures for the Characterization of Vaccines
Vol. 104 Langen, 1999	Development and Clinical Progress of DNA Vaccines
Vol. 105 Paris, 2000	Progress in Polio Eradication: Vaccine Strategies for the End Game
Vol. 106 Rockville, MD, 2000	Evolving Scientific and Regulatory Perspectives on Cell Substrates for Vaccine Development
Vol. 107 London, 2000	The Design and Analysis of Potency Assays for Biotechnology Products
Vol. 108 Langen, 2001	Advances in Transfusion Safety
Vol. 109 Washington, DC, 2000	Biologics 2000 – Comparability of Biotechnology Products
Vol. 110 Palm Cove, 2001	Orphan Vaccines – Bridging the Gap
Vol. 111 Utrecht, 2001	Advancing Science and Elimination of the Use of Laboratory Animals for Development and Control of Vaccines and Hormones
Vol. 112 Bethesda, MD, 2001	Immunogenicity of Therapeutic Biological Products
Vol. 113 Berlin, 2001	Process Validation for Manufacturing of Biologics and Biotechnology Products
Vol. 114 Ames, IA, 2002	Vaccines for OIE List A and Emerging Animal Diseases
Vol. 115 Bergen, 2002	Laboratory Correlates of Immunity to Influenza – A Reassessment,
Vol. 116 Los Angeles, CA, 2003	Development of Therapeutic Cancer Vaccines
Vol. 117 Langen, 2003	Consideration of Alternative Licensing Procedures for Vaccines for Minor Species, Minor Indications and Autogenous/Autologous Products
Vol. 118 Langen, 2003	PDA/EMEA European Virus Safety Forum
Vol. 119 Buenos Aires, 2004	Control of Infectious Animal Diseases by Vaccination
Vol. 120 Bethesda, MD, 2003	Advances in Transfusion Safety
Vol. 121 Bergen, 2003	Progress in Fish Vaccinology
Vol. 122 Bethesda, MD, 2003	State of the Art Analytical Methods for Characterization of Biological Products and Assessment of Comparability
Vol. 123 Rockville, MD, 2004	Vaccine Cell Substrates 2004
Vol. 124 Paris, France, 2005	OIE/FAO International Scientific Conference on Avian Influenza
Vol. 125 Kiev, Ukraine, 2005	First International Conference on Rabies in Europe
Vol. 126 Saint-Malo, France, 2005	New Diagnostic Technology: Applications in Animal Health and Biologics Controls
Vol. 127 Sydney, Australia, 2005	Advances in Transfusion Safety – Volume IV
Vol. 128 Florianopolis, Brazil, 2006	First International Conference of the OIE Reference Laboratories and Collaborating Centres
Vol. 129 Bergen, Norway, 2006	The OIE Global Conference on Aquatic Animal Health
Vol. 130 Verona, 2007	Vaccination: A Tool for the Control of Avian Influenza
Vol.131 Paris; 2007	Towards the Elimination of Rabies in Eurasia
Vol. 132 Paris, 2007	Animal Genomics for Animal Health